UNDER ATTACK

Published by CelebrityPress®, Orlando, FL

CelebrityPress® is a registered trademark.

Printed in the United States of America.

ISBN: 978-0-9966887-0-3
LCCN: 2015948692

This publication is designed to provide accurate and authoritative information with regard to the subject matter covered. It is sold with the understanding that the publisher is not engaged in rendering legal, accounting, or other professional advice. If legal advice or other expert assistance is required, the services of a competent professional should be sought. The opinions expressed by the authors in this book are not endorsed by CelebrityPress® and are the sole responsibility of the authors rendering the opinion.

Most CelebrityPress® titles are available at special quantity discounts for bulk purchases for sales promotions, premiums, fundraising, and educational use. Special versions or book excerpts can also be created to fit specific needs.

For more information, please write:
CelebrityPress®
520 N. Orlando Ave, #2
Winter Park, FL 32789
or call 1.877.261.4930

Visit us online at: www.CelebrityPressPublishing.com

UNDER ATTACK

How to Protect Your Business & Your Bank Account

From Fast-Growing, Ultra-Motivated and

Highly Dangerous CyberCrime Rings

CELEBRITYPRESS®
Winter Park, Florida

CONTENTS

CHAPTER 1

THE ENEMY WITHIN – PROTECTING YOUR FIRM FROM INSIDE THREATS

BY JOHN BIGLIN

It started with a phone call: "This is Bill Sansom and I need help. I received a message that hackers had scrambled all of my files on my computer and unless I pay them $300, I will lose them forever. How can I get my files back?"

Bill was the retiring CEO of a large insurance firm. After a few questions it was determined that his system had been infected with a nasty type of malware, "*ransomware*" in this case, called **Cryptolocker**. At that point, I knew that we would not be able to recover his files, and that the next few steps were more about minimizing the damage caused by Cryptolocker.

The good news was that Bill was working on a laptop he kept at home and was not connected to his firm's network. That limited the damage to the firm's files. We were able to recover files from a backup and had him running again in no time. He was lucky.

Once Cryptolocker gets onto a host computer, it begins its process of scrambling the files on that computer as well as any file folders connected through USB-connected hard drives, network drives and others. The files become encrypted using a typical encryption algorithm and cannot be unencrypted unless a special key is used. Hackers charge about $300 to release the key.

I don't think the total damage of Cryptolocker can ever be calculated, but it is estimated to be over $100 million dollars. New variants of Cryptolocker are being created each month.

Most infections of Cryptolocker are triggered by a user inadvertently clicking on a link or opening an e-mail attachment that silently drops the Cryptolocker code onto their computer. In some cases, it might be some other malware that lands on the user's computer which then pulls Cryptolocker in.

Years ago, savvy employees knew to stay away from questionable websites or weird looking e-mails/attachments. Today, many malicious e-mails and attachments look valid, and opening them triggers the infection through an honest, and unknown, mistake.

We received a similar call from a Credit Union, which became infected despite having its yearly external IT Scan completed by its auditors. Company executives assume that passing external security scans and penetration tests means their firm is fully protected. While these scans and tests are extremely important, there are just as many threats that emerge from the *inside* of a company.

Much has been covered in the media about hackers from other countries attacking corporate networks and computer systems. In response, many firms focus on strengthening their network security by installing advanced firewall devices, intrusion detection / prevention systems and countless other tools. Then they have vendors run external vulnerability scans on their network to check for weak points where hackers could gain access to their systems. All good steps.

Very often, business executives do not realize how many data breaches and data losses occur as a result of someone *inside* the firm doing something that allows the bad guys into the network.

Some of these internal threats develop through intentional actions by internal employees and contractors working inside the firm, but more often the vulnerabilities are created *unintentionally* by them.

Phishing scams are probably one of the most common entryways for cyber-criminals. In a Phishing e-mail, the sender is masquerading as someone else and they typically try to get you to open an attachment, click a link, or something similar. That action usually attempts to download some type of malware-related files, even if it doesn't actually do immediate damage it can remain dormant. In some cases, it brings you to a familiar-looking website and asks for your login credentials – for the sole purpose of acquiring your ID and Password for use later on. Criminals have become more sophisticated and can send e-mails that look like they are from FedEx, American Express, Verizon and countless other legitimate firms.

Even worse is "Spear-Phishing", in which someone uses public or other available information in an attempt to target an individual user specifically by acting as someone they know by sending an e-mail.

We had a case in which a firm's controller received an e-mail from Sean, the CFO. The e-mail read:

"Mary, I forgot to ask you to send a wire transfer on Friday and the vendor never received the funds. Would you please wire $43,822 to them using the following wire transfer instructions:

<wire instructions redacted>

Thanks,
Sean

Mary responded:

Sean,

OK, I can prepare the wire, but I can also FedEx them a check and it could be there tomorrow. Let me know your preference.

Mary

Sean responded:

Mary,

Thanks, but I needed to get this to them Friday and I am embarrassed that I forgot. Can you get this out right away?

Thanks,

Sean

Mary said 'Sure' and proceeded with filling out the Wire Transfer form. Fortunately, Mary knew Sean was in the office and mentioned the wire transfer request to him in-person. Instantly, they knew something was wrong.

Upon further investigation, they found that the crooks had gone to the trouble of registering a very similar domain name "BalamoreBio.com" instead of BallamoreBio.com (name changed to protect the company's privacy) with only one 'l' character instead of two – a very tough change to spot. Sean@ballamorebio.com looks almost identical to Sean@balamorebio.com.

So the crooks knew important details such as the names of the CFO and Controller, and they were believable in their communication.

We had a similar incident at our firm. One of our Sales Reps received an RFP (Request For Proposal) from Rutgers University for a project which included upgrading some equipment and reworking a data center. After conversing with the Rutgers representative a few times, he responded to the RFP with a proposal. A week later we were notified that we had won the project but needed to ship the hardware ASAP and have it delivered prior to the project planning meeting. Much of the correspondence was mainly via e-mail, but my rep had at least three or four phone calls with the so-called purchasing manager from Rutgers.

As the hardware orders were being prepped, someone noticed that the address was not at a valid Rutgers campus – instead it was in a trailer park! Then we looked at the Purchase Order in more detail. That's when we realized the person he was communicating with was *jbrennan@ rutgers-edu.us* when his address should have ended in *rutgers.edu* or something similar.

No external scans of your network are going to trap for these issues. We can install high-quality firewalls and all kinds of protection software, but that will not catch everything. What you *can* do is train your users how to deal with links they see and e-mails they receive. Information Security Awareness Training should be an important and required part of a company's regular employee training and orientation.

While there were many reasons for the Target breach, one key factor that was reported was that one of Target's HVAC vendor's office computers was compromised and that vendor had access to the Target Network. Through a phishing e-mail, the bad guys were able to put malware on the HVAC vendor's computer and it captured the credentials the user entered when they accessed Target's systems. When the bad guys wanted to enter Target's network, all they needed to do was login using the HVAC vendor's computer and credentials (which the malware had control over) and then drop their own programs on Target's Point of Sale systems. No alarms were set off because there was a *valid* user name and password connecting to the network.

That HVAC vendor had anti-malware software on their PCs, but it was reported that they were using the *free* version and not the *paid* version. One key difference between the two? The *free* version did not monitor the PC consistently, and only scanned for Malware when the user would run a scan. The *paid* version watched for Malware at all times.

Our firm has countless stories of real-world hacks that have been triggered from the inside of a firm, not from the outside. Listed below are some of the types of vulnerabilities that create threats from the inside:

TYPES OF VULNERABILITIES CREATED FROM THE INSIDE OF A FIRM

- Users clicking on website links and email attachments that are triggers for installing malware, viruses, etc.

- Users taking a laptop home and letting someone use their laptop who accidentally installs malware or viruses by downloading free movies, apps, music and other files from non-official sources.

- IT staff or contractors making changes on key systems or firewalls while they are trying to install or upgrade an application, but leave the system or firewall in an unsecure state.

- Weak passwords. Too many people use weak, easy-to-guess passwords across several systems that they use.

- Shared passwords. When multiple people in a firm use the same account and password.

- Poor controls of administrator accounts. We recently conducted an assessment of a firm that had three system administrators but had over a dozen active administrator login accounts! We found that some were for former employees or consultants. Those additional nine active administrator accounts each were an unnecessary risk to the firm.

- Ignoring security patches and updates that are needed to keep each computer protected.

- No real-time anti-virus or anti-malware software in use on all systems and servers.

- No real-time anti-spam to trap for bogus e-mail.

- Having a firewall in place without checking to ensure it is continually updated.

A HUGE FACTOR: EMPLOYEE AWARENESS

Our firm provides a special security awareness seminar for client employee groups to educate them on many of the scams that are out there and how to avoid them. In every session we have done, we have had countless client staff members tell us they never realized how easily it would be for them to accidentally create a security issue for their firm.

NINE BASIC STEPS TO TAKE TO PROTECT YOUR FIRM FROM INTERNAL THREATS

Below are nine relatively basic, simple and low-cost steps you can take to protect your firm from many types of internal threats:

1. Backup your systems and data locally and offsite, and test the restore process. There are some fantastic low-cost solutions available for backing up servers and user / workstation data offsite on a daily (or more frequent) basis.

2. Install high-quality firewalls at every office/site that has access to the Internet. Most firms have firewalls in place, but in more than 25% that we assess, we find that their firewalls have not been updated in months or years with the latest patches and updates. Have an IT partner check your firewalls regularly, even if you have someone doing it internally.

3. Enforce complex passwords on your network, desktops/laptops, tablets and mobile devices and require that they are changed at least every 180 days, preferably every 90 days.

4. Maintain separate Administrator-level accounts for servers and user computers such that no one uses a system on a day-to-day basis logged in as your domain Administrator. Instead, they should log in as a 'Standard' user that doesn't have Administrator privileges. That

way, if they accidentally download any malware, it cannot do as much damage if they are logged in as a non-Administrator level user.

5. Think before you Click. Links in e-mails and web pages can be masked or spoofed whereby it looks like a valid link but it really takes you to a fake site that will either try to capture your login/password or cause some other type of damage to your system (malware, etc.). Show all your users how to hover the mouse pointer over a link in an e-mail before clicking on it. Windows and Mac OS computers will show where the link actually goes to. For any sort of Banking, Financial, Utility, or Healthcare-related link, we recommend telling users to simply open a new browser and type the URL/web address into the browser rather than clicking a link in an e-mail. It takes about 60 seconds longer, but can save tons of time, money, stress and distraction by preventing a malicious event from occurring.

6. Absolutely ensure that a real, high-grade Antivirus/AntiMalware (AV/AM) system is installed on all computers in your network and that it is regularly monitored and updated centrally. When conducting our IT Assessments for new clients, we often find that there are some, or many, systems that either are not running AV/AM or that it is significantly out-of-date. Also, the first thing many malware programs try to do is disable the AV/AM software running on the computers it is infecting. By staying vigilant about keeping the AV/AM running and up-to-date, it will help you keep every system protected and will likely alert you if something is wrong.

7. Keep all systems up-to-date with the proper security patches from Microsoft, Apple, Adobe and others. Most modern systems today regularly check for available patches, but we find many users decline them or put them off for weeks or months. There are ways to set your systems to automatically patch as needed, although some clients of ours prefer to plan and communicate when patches are to occur. Regardless, it is important to constantly monitor and report on your patch levels and to deal with critical patches relatively quickly after they are released.

8. Do not allow open, unencrypted WiFi networks in your office. If you must have a guest wireless network, be sure to use encryption and a password, and change it regularly.

9. Try not to use open, unencrypted, free WiFi networks when traveling – such as in coffee shops, etc. My team can easily capture user credentials and other information with some basic tools when a WiFi network is open like that. The bad guys do it all the time. In some of the speaking engagements I have had, I have put up a phony "Free WiFi" network with no password and I see many people connect to it and try to use it. If I wanted to, I could capture a lot of their information.

CONCLUSION

I have included only a handful of the vulnerabilities that emanate from the inside of an organization. In 95% of the cases, the crooks are setting things up from the outside, but the threats are often enabled, albeit accidentally, due to what happens on the inside. This chapter was not intended to be a "doom-and-gloom" chapter, but I am hoping that the information I shared gave you a sense of some of the risks that exist in businesses across America and around the world that are often triggered from the inside of a firm, regardless of the strong external security and scanning that is in place.

About John

John Biglin is a speaker, best-selling author, seasoned business executive and board-level advisor who leverages his 27 years of technology industry experience to help senior executives establish the ideal balance between performance, reliability, security, compliance and cost. Using a "Business-First" philosophy, John focuses on solving business needs rather than simply prescribing a particular technology.

John has a long history of launching, developing and managing several successful multimillion-dollar firms, while helping other business owners do the same.

John is currently the *Chief Executive Officer* of Interphase Systems, Inc., a premier Technology and Managed Services firm serving the Life Sciences, Financial Services, Professional Services, Healthcare, Manufacturing, and other industries. Interphase Systems was founded in 1995 and has offices in Pennsylvania and New Jersey.

After 20 years of working directly with Life Sciences firms, John pioneered the creation of *Ready-IT BioPharma*™, the Life Science Industry's first turnkey IT Managed Services Platform that integrates core business and IT systems combined with support services, disaster recovery and regulatory compliance. His firm has since expanded the Ready-IT™ concept to other industries and integrated Cloud technologies into the service.

John's prior experience includes working as a senior consultant with IBM Corporation as well as other IT leadership positions. He has also been an adjunct professor of Business at Gwynedd Mercy College in Southeastern Pennsylvania.

John is regularly interviewed by major industry trade publications *(PharmaVoice, Pharmaceutical Commerce, eWeek, InfoWorld, Technically Philly*, etc.) and has been a speaker at many industry conferences, including:

- Virtualization Exchange (Keynote Speaker)
- Virtualization Technology Conference
- Philadelphia Cloud Computing Conference
- Financial Managers Society
- Pennsylvania Bankers Association

John holds a Bachelor's degree in Computer Science from Rutgers University and an MBA from Penn State University.

CHAPTER 2

THE FOUR MYTHS OF WEBSITE SECURITY

BY BILL THOUSAND

Sometime in 2001, our firm received a call from a potential client. They found us through an Internet search, which landed them on our website. We'd never met or even heard of his business before, but he was only about 20 minutes away and he seemed genuine and in need, so we offered to set up a meeting immediately.

He explained that he wanted to sell audio content from his website. Nothing risqué, just going from physical media sales to instant web delivery. Everything he wanted was within our capabilities, so we thought we could help him out.

The meeting was going well, until he looks at us and says, "The most important thing for this site is that in really big bold letters on the top of the page I want it to say, **'If you steal my stuff and put it on Napster, I will find you and I will @&$#! kill you!'** "

It was like one of those movies in which you hear a needle scratch over a vinyl record.

Of course he was kidding, right? To the contrary, he was stone-cold serious. We suggested to him that an open-ended threat of physical violence may discourage customers, but he could not be dissuaded. He really wanted that message on the site, so we wrapped up the meeting and let him know we would be in touch.

We did not end up working with this client, but ever since then, anytime anyone asks me about how to prevent hackers from breaking into their site, I can't help but muse about that idea. Not a good or legal one, but nevertheless, it was thinking outside the box to try to mitigate theft.

If you live in the real world like me, threats are not an option, but neither is cyber crime. If you are a business owner, marketing manager or anyone in charge of your organization's website, there are some legitimate things you should think about in terms of website security.

We are often surprised by the lack of attention security gets in terms of website planning and implementation. Some myths exist that prevent smart business owners from making the best decisions for their cyber security. I have listed the top four below in hopes of dispelling them, and changing the misconceptions that continue to allow cyber criminals from stealing money and data, from legitimate business owners.

THE FOUR MYTHS OF WEBSITE SECURITY

<u>Myth 1</u>: *I'm too small, therefore I'm not a target for hackers and cyber criminals.*

<u>Reality</u>: *False!*

We often hear folks say they aren't worried about their website being hacked because they are too small to be worth any attention.

This couldn't be further from the truth.

Small and medium-sized organizations are a HUGE target! These organizations are the 'low hanging fruit' to hackers. Small companies often have fewer resources and very often pay less attention to security, making them easy targets.

We recently had a small business move their website to our data center. We copied the entire site from the previous hosting provider as-is. The reason for the move was performance concerns and lack of support.

The only change we made to the site was pushing all site-generated e-mail through our outgoing e-mail gateway. We require this for all the sites we host.

Within the first few days of putting the site live, our alarms went off showing that our data center was seeing a massive increase in undeliverable e-mail.

After further investigation it turns out the site had been compromised months earlier and had been sending SPAM on behalf of some seedy organizations.

This is exactly why hackers love smaller businesses. This company was using a high volume, low support hosting provider and had no way to know they were hacked.

<u>Myth 2:</u> *I'm safe, because I use sites like WordPress, Drupal or another CMS.*

<u>Reality:</u> *False!*

First, almost every website on the Internet today is utilizing some kind of Content Management System (CMS) as its engine or framework. These systems allow you to add, edit and remove content from your website in an organized manner, often without knowing any programming.

Even if you have no technical knowledge, you should know what CMS your website is built upon and why it was chosen for your organization.

Secondly, from simple brochure sites to complex e-commerce deployments, many websites are built using an Open Source CMS.

Open Source software is community-supported software. Developers, sometimes from all over the globe, contribute ideas, code and support.

While Open Source components are found in everything these days from TVs to toasters, the Internet is where it started and web development is dominated by Open Source projects.

WordPress is the most commonly used Open Source CMS platform, especially for smaller organizations, but the concerns are the same regardless of the system.

Before I continue, I think it's important to clarify that Open Source as a concept is GOOD! In fact, it's GREAT! The Internet itself was built on this philosophy.

I am not suggesting that WordPress or any of these platforms are bad, in fact many of them are very powerful and tremendously useful. I think it's important that you understand that by its very nature Open Source is inherently insecure since everyone, including hackers, have access to the source code. Hackers are constantly analyzing the source code of these projects looking for weaknesses and vulnerabilities.

If your web developer or designer tells you that WordPress or any other CMS is secure and not to worry, I would suggest worrying about your designer or developer!

So, if the question becomes, should I use an Open Source CMS? The answer is that you're asking the wrong question. If your designer or developer is intimately familiar with the platform you are considering, including the security risks, an Open Source CMS can be a very safe and effective solution. On the other hand if your designer or developer is cavalier or seems unaware of the security risks involved in these platforms, I would consider looking harder at your choice.

Some questions you can ask your designer or developer to help assess his or her ability to manage your website in a secure manor:

1. What Content Management System are we using or are you recommending?
2. Why are we choosing that platform?
3. What are the security risks?
4. How are we planning to mitigate the security risks?

A good designer or developer should be able to provide clear and concise answers that give you a strong sense of comfort. The answer to the last question should include things like regular updates, regular backups, change control, monitoring, vetting of third-party plugins and testing processes.

<u>Myth 3:</u> *It doesn't matter if my site becomes hacked.*

<u>Reality:</u> *Huh?*

"We don't have to worry about security. We backup the site every day. We can just restore it if it gets hacked; no harm, no foul!"

"Our business doesn't depend on our website, so if it gets hacked, it won't hurt our bottom line!"

These are common responses I hear, and it may be unbelievable that this attitude actually exists, but we've run into it, more than once even.

Business owners, marketing managers, etc. think that backups alone are enough to mitigate the risks of hackers or that it really doesn't matter either way because it won't cause any loss of business if the website is turned upside down.

A newer client recently called us because their e-mail was suddenly being blocked. Their website was a brochure style site that didn't really generate much business, but like most organizations today, they use e-mail for everything.

After a few hours of investigation, it was discovered that their domain name had been blacklisted due to their website being compromised and used by hackers. Since their website was built by an inexperienced designer and hosted at a discount-hosting provider, they didn't even have useful backups of their site to restore from. Three days and many consulting dollars later, e-mail started flowing again and the site was rebuilt from some old source files.

This example is only the tip of the iceberg in terms of the damage a hacked website can do to your business. The damage can be exponentially worse if your website is used for ecommerce or connected to your organization's data in some way and customer information or other confidential records are stolen and/or exposed. Many states have laws on the books now that require businesses to notify customers if their personal information has been stolen or exposed.

Be aware of the potential security risks and perform due diligence. A little attention to the security of your website can go a long way in terms of prevention and recoverability.

Myth 4: *There is nothing you can do to prevent hackers.*

Reality: *True and False!*

It is true that you cannot guarantee 100% success in preventing hackers from compromising your website. These cyber criminals constantly change their tactics. We work constantly every day to keep up with new vulnerabilities. Consider though, cyber criminals are well-funded by their exploits. They are dumping all their time and resources into their craft.

There are many steps you can take to minimize the risk and the damage.

PREVENTION

Foremost, choose a good designer and developer. I cannot stress enough how important it is to have a good team managing your website. This team should include at least one of BOTH a designer and a developer. It's very rare to find a single person good at everything. If you only have a designer, find a developer, even on a contract basis, to make sure the site is implemented properly and securely.

If you're not sure of the difference, a web designer is someone good with graphics and aesthetics. Their tools of the trade are Photoshop, Illustrator, HTML and CSS. A developer is a programmer. Someone who knows the core programming language upon which your site's framework is built.

Ask questions of your developer to establish the proper level of comfort. A good developer should give you confidence they are technically savvy and be able to explain things in very simple but clear terminology. Consider it a red flag if his or her answers are vague or dismissive when it comes to security. If you're using outside help to build and maintain your site, make sure your organization is a good fit for the vendor you choose.

You should also have a good support model. What that means is everyone from your designer and developer to your hosting provider should be ready and able to help if you have security issues.

Here are some things your team should consider and create plans to address:

1. *Monitoring:* Use an automated system to make sure your site is up and operational at all times. Many modern monitoring systems can check the site content to make sure key words or phrases are present and alert you if the site is redirected or dramatically altered.

2. *Hosting:* Make sure your site is hosted at a reputable hosting company that employs industry standard security precautions such as anti-virus protection and regular operating system updates. Many of the discount hosting providers are huge targets for hackers because the types of sites hosted are usually small and have weak security. Consider a low volume hosting provider to help reduce this risk.

3. *Platform and Plug-Ins:* If your site uses an Open Source CMS or components, be sure the organization behind these components is reputable, and make sure everything is being updated on a timely basis.

4. *Change control:* Some sites are so dynamic that it's hard to implement a change control protocol, but ask your developer if it's possible to track changes and send alerts in the case that any files that are not supposed to change are changed. This is a strong indication of a hacker.

5. *Penetration Testing:* If your site is subject to PCI compliance testing, your credit card provider may be doing this for you. Otherwise, find a third-party or consultant to conduct a penetration test on your website. This is something we do often for our customers. It involves running an automated battery of tests against your site for known vulnerabilities.

PREPARATION

In addition to trying to prevent hackers, being prepared to deal with them is equally important.

Some areas of preparation to consider are:

Backups: Backups are important as your last line of defense if your site gets hacked.

Support: Make sure you know who to call when you have an issue. Surprisingly, we run into organizations on a regular basis that don't know who hosts their website.

Exposure: Make sure you know what company data is connected to your site. Many times connections are made by your company databases to your website. You should be aware of these connections so you know what kind of exposure you are risking.

Consider these myths, and how you can use the information I have provided to begin building a strong security infrastructure within your business. The time, effort and even money used to produce such an infrastructure will be worth ten times more if you can keep your company safe and less vulnerable to hackers and cyber criminals.

About Bill

Bill Thousand has been working in the IT industry for over 25 years. Bill started writing software on his C64 at the age of 12 and by high school was providing computer help and programming small projects for nearby businesses. In the early 1990s, he was a partner in Professional Computer Solutions, Inc., which grew to one of the biggest Microsoft Dynamics partners in the Midwest. During that time, he obtained his Novell CNE, Microsoft MCSE and a variety of other industry certifications. Wanting to focus more on IT support and custom development, Bill founded **Clarity Technology Group, Inc.** in 1999 and has been providing IT support services, IT security consulting and web applications development ever since.

Bill is both an experienced programmer and an IT support engineer, and as such is uniquely qualified to provide security consulting services. Understanding both technology as well as how the hacker's mind works provides a much broader view of the issues and risks.

Bill's philosophy on providing IT support is that you need to really understand your customer's needs before you can help them. He trains his staff to ask questions and pay attention. "Before you can build a solution, you need to know why you're building it, who will be using it, when they will be using it and how they will be using it." Bill's goal is to create a long-term trust with his clients that can only be obtained by really understanding what they are trying to accomplish.

When not working in the IT world, Bill spends much of his free time in ice rinks all over the country watching his three kids play hockey.

You can connect with Bill at:
bill@claritytech.com
https://www.linkedin.com/in/billthousand

CHAPTER 3

BREACH PREVENTION

BY BRANDON JONES
- Attitude Computers

In late 2013, we had a call from a local small business in the area, as one of my technicians listened to her trembling voice over the phone, it was shear panic about to break through. Something had happened, and all her instincts told her it was seriously bad! Consider that this was Monday morning at 9:30 am, and after a nice weekend with her family, Linda was back in the office and really losing it.

Linda runs a local marketing company that provides targeted data-driven marketing plans and deliverables for other small businesses. When she arrived at the office it was business as usual - get the coffee on, water the plants, open the blinds in the main lobby - you get the idea. It was her business and she liked being the first one in at the start of the week. It gave her a chance to clear her mind and prepare for the crazy, busy week ahead. It was the only time she didn't have to answer employee questions or solve some unsolvable problem for a certain staff member (you know the one) that she'd previously solved for them the last three weeks running.

As she poured the cream in her cup of steaming hot coffee, she began mentally planning out her day. First, she had a very large deal to work on; it was a contract with a company that would equal half her client base in size. She had a conference call with them at 1:00 pm to finalize all the terms and set a schedule for the deliverables in their agreement. After that she had payroll – she had smartly chosen to pay her staff semi-monthly (so she never had to deal with a 3-pay-period month) and it was the 15th, so payroll had to go out today. She decided she would spend the rest of the morning preparing for her extremely important call with her potential client.

Linda sat down, and took a sip of coffee before logging into her computer. First, she opened Outlook to run in the background so she could stay on top of an important messages that came in while she worked. Next, she went to open the excel spreadsheet with the breakdowns for the meeting this afternoon.

"Huh," she thought, "that's a strange error message." The screen displayed – *There's no program associated with this file?*

"What does that mean?" she pondered.

She tried another Excel file - same message. Maybe a word document? Nope, that doesn't work either. She knew the server was running, she could see it was on, and all the folders were in the right place. What the heck was going on?

It was now just before 8:00 am and her staff was slowly filtering into the office. She called one of her colleagues, Brianna, "Can you check your computer and see if you are having the same problem?"

"Sure thing, Linda, let me just finish logging in, ok?" And in a moment she replied to Linda, "Huh, I can't open anything either? This is strange." Suddenly the office came alive with similar complaints and exclamations. One of the other workers. John, had something different on his screen – *Deadline lapsed, and files encrypted.*

"This appeared on Thursday, but I figured it would go away over the weekend," John explained.

Linda began calling around in a panic and had tried to call a number of other IT companies but only got a voicemail, until she finally connected with us. Linda knew she was in trouble and I believe the main part of her call was to hear someone on the other end of the phone say, "It's going to be all right, we see this kind of thing all the time, all we have to do is such and such and it'll all be fixed."

We discovered that the huge problem that Linda's company was experiencing could have been prevented. It turns out John was surfing the net on some unsavory sites that previous Thursday. I am sure you get the picture. And he picked up a virus that encrypted all their files on the server unless they paid the ransom requested.

What neither John nor Linda knew is that there is a whole criminal underground that is looking for ways to monetize the lack of security most small businesses have in place. Organized crime has figured out that by getting out of the "old game" of street crime like theft, drugs and prostitution – they can actually make more money. The best part is that they don't have to be there in person to commit the crime, which allows thcm a grcatcr chancc to gct away with it.

Most small businesses don't even know they're a target. These cyber criminals get in through their bank accounts, Internet connections, cloud services, servers, desktops, laptops, employees, USB drives, wireless, smart phones. There are virtually hundreds of ways Cyber Criminals can get in. It comes down to simple math – Cyber Criminals only need to find just one way in, while conversely, Cyber Security professionals must block every conceivable point of entry. Even with this sobering fact, most small business networks, aren't patched up-to-date, they don't have good policies in place for employee usage, or adequate backup systems. Even if they have a backup system, it has become "common practice" to set a backup system up once and never check it again. Companies and their IT professionals cannot be sure the system is backing up the right stuff because they never test it.

How easy is it to become infected? Even bright people can be duped. In a study, a USB drive was left in the parking lot of a company. 50% of the time an employee walking past would see it, pick it up and plug it into their work terminal to see what was on it. In this case an employee accidently circumvented much of their security with a simple mistake - because once plugged in the flash drive automatically installed a RAT (Remote Access Tool), that could allow someone full access to the network as though they were in the building. Further into the study, the same USB drive was left, except this time it had a company logo on it. In this case, the unsuspecting employee would plug it into their work computer an astounding 80% of the time - and get this, it didn't even need to have their own company's logo - it could be McDonalds, Wal-Mart, or anything else, it really didn't matter.

If you're wondering how Linda made out in the end? Don't worry, we were able to get all of her files back to the way they were – but it took a couple days. Since that Monday – and I believe Linda still refers to that day as "Black Monday" – all the stops have been put in place to prevent something like this from ever happening to Linda again. After getting her back up and running, Linda agreed to let us do a Security Assessment on her network to find problems and fix them. We went to work straight away, and found many holes in her Network's Security.

From there, we put in place effective protection against Viruses, Malware and Spyware. We upgraded her network, put in proper firewalls and SPAM protection, we even set policies in place for her staff – so they'd know what was acceptable and what wasn't.

Once the basics were put in place, we then went to work on making sure Linda's staff understood the importance of <u>NOT</u> having the word "password" or "Pa55word!" as the only thing stopping the World Wide Cyber Criminals from getting at their most important data.

Then we moved on to more advanced items, such as giving Linda full reporting on her staff's online activities. Now she knows, that when someone's report doesn't get done on time, it could have something to do with the nine hours they spent on Facebook last week. In addition, she's even empowered to block certain websites if she deems them a waste of her employee's time. These changes, coupled with computers that are now running as fast as the day they bought them, have increased her staff's productivity. Linda's staff became so effective that she didn't even need to hire new employees to support her big new client! And yes, she still landed it!

After spending some time with Linda and working together, we found out that she hadn't taken a vacation since starting the business 12 years ago. Digging further, it seemed to be due to her fear that something would go wrong at the office and she wouldn't be there to help solve it. Linda's fear was real, even if the likelihood of a major business-destroying event happening while she was away was very low. In answer to this, we configured remote access to anything she needed at work. She could remotely access anything from her smart phone, laptop or tablet – talk about making it easy to work from the beach – fair warning it's not for everyone. And with today's roaming plans, and WIFI available at nearly every resort – this was a no brainer – and boy was her husband excited to finally get away!

Upon reflection, the biggest change for Linda is the conversation that we now have with her. It's no longer about dealing with slow, failing, weak technology, server crashes, viruses and other drama-inducing problems. It's now about putting technology and systems in place to support her business and its growth. The conversation is now about getting to the "Next Level" instead of running around putting out fires.

If you are ready to protect your company, investments and your customers, then do some research, ask around, and find a few really good IT Security Companies that can handle your security needs. This does not always need to be done in-house. You can hire an outside firm to do all of your initial heavy lifting to secure your systems and they can also monitor and ensure that your system remains secure against outside threats. Make sure they have excellent references and check them out through sources such as the Better Business Bureau or even Yelp! Then have a few of your top choices come out and do a free Risk Assessment of your Network Security. Next step is easy, decide who you like best and hire them. The only thing you would regret is not making the decision of hiring someone to secure your systems. Ask Linda what she regrets.

I know some of you automatically flip to the back of the chapter for some quick takeaways that you can implement right now—and I don't want to disappoint, so here they are:

QUICK TIPS, THAT YOU OR YOUR IT PERSON CAN DO RIGHT NOW

Backups: In a couple words, TEST them. Do a test restore right now to see what your backup system has actually been backing up all this time. I don't know how many times a backup system is setup and forgotten, and is only tested when something goes wrong. That's the worst time to find out that it hasn't been backing up the files you need – and by then it's often too late.

Create an Acceptable Use Policy: Make sure your staff knows what is and isn't allowed on work computers, and on company time. This is a good first step to getting more effectiveness from your people - remember John? Maybe they just don't know.

Create a Password Policy: Make sure your staff knows what a strong password is, and that they are expected to use them. Make it clear that in the event of a Network Breach, they will be held accountable if it was caused by them using their pets name "snuggles" as a password. For more help on this, a quick Google search will reveal current best practices.

Software Updates and Patches: First, DO THEM. Every time Microsoft releases an update, its normally because people got hacked, their data got stolen, their business went under, or their identity compromised. By the time Microsoft releases the update to prevent it, years may have gone by, and thousands more people affected. Don't suffer the same fate by not updating your computer.

Antivirus: First, make sure you have a good one. Second, make sure your subscription is up to date. Third, make sure it's automatically scanning on a regular basis. And last, be sure it is automatically downloading the latest updates. Remember an Antivirus program is a program, and can only protect you from what it is programmed to protect you from. This means that when a new virus comes out, you are not protected, not until the Antivirus manufacture notices that a new virus is out, then writes an update to protect you from it, then allows you to download the update. This can take a few weeks, so there's always a gap in protection from a new virus. But if you don't get the update, you'll never be protected.

About Brandon

Brandon Jones helps his clients to elevate their IT from a "Problem Center" into a "Profit Center". He does this by first working out all the kinks in the current system, followed by putting systems in place to allow IT to drive efficiencies in the business.

Brandon's passion with computers began when he was in third grade and his parents purchased a computer for him that Christmas. Prior to this he was always taking apart anything electronic or mechanical that he could get his hands on. He spent most of his time building small electronic devices, and computers seemed like the next step; he built his first computer by seventh grade.

By High School he was coding and wrote several games, one was based on Pong, and another was akin to Space Invaders but with full graphics. During his post-secondary period, he was hired at an IT company doing computer repair and supporting local businesses. Brandon always strove to find a better way to do things, and constantly challenged the *status quo*. Eventually, Brandon became disenchanted with working for someone else.

In 2003, Brandon started his own IT company, **Attitude Computers**. In the beginning he mostly did repairs for friends and family. Eventually he had to hire staff to help support the hundreds of new clients flocking his way. As the company matured, so did his client base, and he now only supports business clients, and always in the best way possible.

If you're interested in finding out more on how to find a good IT Cyber Security company, please email him at: brandon.jones@attitudecomputers.com with the words "Cyber Security Free Report" in the subject and he'll be happy to send you a report on what to look for, what to watch out for, and what to expect when trying to find a Cyber Security company you can trust.

CHAPTER 4

CYBER INSURANCE – WHAT IS IT, AND DO YOU NEED IT?

BY DEREK DAVIS

On a Thursday afternoon, an accounts payable clerk in your company has just completed the weekly task of transferring funds from your bank account into a separate payables account. She does this every week after cutting checks and getting a payables balance. Your accounting department considers this a good practice, because it doesn't allow anyone to fraudulently transfer money from the payables account, but on this Thursday afternoon, your CFO gets a call from an officer at your bank.

Mysteriously, a large sum of money was transferred to a foreign bank at the same time your employee sent the funds to the payables account. After some digging around, the bank transfers you to their fraud department, where a computer forensics expert asks to check your payables computers for a virus. Of course you have antivirus software on your computer, so that's not possible. Your bank makes your clerk use a Secure-ID card that changes every sixty seconds. You are certain you haven't been a victim, but your pulse quickens.

After running a few checks, the bank's forensics expert produces a report that shows your payables clerk's computer is infected with a "key Logger" program that allowed someone in an Eastern European nation to "see" the account information as well as the Secure-ID card data your clerk entered. The transfer was authorized to send $75,000 to an account in Belarus. *Your money is gone*.

Welcome to the world of cybercrime. After a gut-wrenching afternoon, you pick up the phone and call your insurance agent – your last glimmer of hope – to file a claim.

WHAT IS CYBER INSURANCE?

Despite every feasible effort made by individuals and corporations to combat cybercrime, the fact remains that machines are still vulnerable. Dr. Vinton Cerf is credited with designing key building blocks of the Internet in the 1970's and 1980's. It's been said that Dr. Cerf and the group that developed the Internet could have somehow prevented today's insecurity, but blaming the technology is like blaming urban planners for muggings. It is true, however, to say we live in a much more inter-connected world today, with many more threats, than Dr. Cerf envisioned in the late 1970's.

While the security industry struggles to keep up with the challenges of cyber security, the insurance industry has identified an opportunity to provide a mechanism to help offset the potential losses from this type of activity. **"Cyber Insurance"** is the name given to this type of product. Your first thought may be that insurance companies are only trying to monetize this threat, and though they are indeed trying to profit– they are also providing a necessary product and forcing necessary changes.

Generally speaking, Cyber Insurance is designed to help mitigate loses resulting from a variety of computer-related security issues, including data breaches, business interruption, loss of private information, and network damages.

Cyber Insurance is a market worth taking notice of, as it can both protect businesses and help them recover if losses do occur. The Department of Homeland Security believes that a competitive and robust "Cyber Insurance" market can help reduce the number and the seriousness of electronic attacks by doing two things: encouraging firms to make use of preventative measures in order to receive more coverage or better rates and offering insurance premiums based on a firm's level of self-protection.

Each layer of protection a company utilizes to help prevent cybercrime should also help the company receive better insurance coverage and lower rates. Conversely, a company that ignores the possibility of a cyber-threat and does not implement good security practices could pay very high premiums, or be declared un-insurable.

WHAT CYBERCRIME RISKS DO BUSINESSES FACE?

In today's business world there are many cybercrime risks and a variety of ideas and solutions to help offset or prevent these occurrences. Several primary risks face companies large and small:

1. **Physical security** can be compromised in many ways, by employees bringing their personal computers into the office, loss or theft of company computer equipment, and the exploding use of mobile devices.

2. **Data** shared on social media sites can be used to piece together enough information to enable hackers to target individuals, facilities, and organizations.

3. **Loss of information** by hacking into companies that store credit card data, health records, or other personal data is an all too familiar threat which has already cost large corporations millions of dollars.

4. **Companies face ongoing and escalating threats** from viruses and malware, email fraud, compromised web sites, and other Internet-based threats.

HOW DOES CYBER INSURANCE
PROTECT BUSINESSES?

Cyber Insurance policies are designed specifically to include coverage for these types of events. It is important to note cybercrime is not usually covered by a company's General Liability or Professional insurance policies. As early as 2001, language began appearing in General Liability policies to exclude coverage for the loss of electronic data.

Cyber Insurance policies provide liability coverage for security breaches and losses of confidential information due to unauthorized access to systems. Cyber Insurance policies also help cover costs associated with consumer notification, customer support, and even providing credit monitoring to affected customers.

Many direct and indirect costs a victimized company faces, including the following, can be covered by a Cyber Insurance policy:

- Restoring, updating, or replacing business assets
- Business interruption and security consulting
- Cyber Extortion or Cyber Terrorism
- Regulatory Compliance issues
- Liability, Slander, Copyright Infringement, Product Disparagement, etc.

I DON'T NEED CYBER INSURANCE... DO I?

In my business, we preach to our customers the importance of backing up data. As a seasoned Managed Services Provider, we know eventually something is going to happen to a customer's computer to potentially cause a loss of data.

The Gartner Group says that only 6 percent of companies survive longer than two years after a major loss of data. Further, the group states that 43% of companies were immediately put out of business by a "major loss" of data, and another 51% permanently within two years.

We equate the risk of losing your data due to a computer crash to the risk of losing data due to cybercrime, viruses, malware, etc. **If your data is worth backing up (which it is), then Cyber Insurance is worth considering.**

As a trusted technology partner to many local companies, it is worth noting that even my company was not aware of this type of insurance until recently. As part of our own Professional Liability and Errors & Omissions Insurance renewal process, we were approached by our carrier to look into taking out additional coverage for our company. While we contemplated whether or not this insurance makes sense for us, we realized we need to evaluate several things.

HOW DO I EVALUATE CYBER INSURANCE?

There are a number of questions to consider when looking at Cyber Insurance. Based on the complexity of how your data is managed, here are some things to consider:

"The Cloud"

A large amount of data is now stored in data centers around the world generally known as "The Cloud". In many cases, companies do not have a personal relationship with the people that run these data centers. Is there a contingency plan to continue operating in the event this environment is unavailable for several days?

If you use "Cloud Storage," you must determine if your Cyber insurance would pay for costs associated with cybersecurity issues arising from the use of the Cloud. Would the insurance policy help recover lost data or mediate between your business and a "Cloud Storage" Company?

Email Issues

Many companies have their email hosted on third-party email systems. Microsoft's "Outlook.com", Google's "Gmail", plus many other companies deliver hosted email on a platform running a high-end email system. Having your email hosted outside the building is a good idea for many companies, but creates dependence on these hosting firms to deliver, store, and back up your email data. Your Cyber Insurance Policy should address contingency planning in the event some piece of your email delivery system fails.

Types of data

Analyzing the type of data that your company handles is a critical step in determining your exposure to a cyber-event. If you handle financial information, you may already be required under FINRA or the SEC to have a number of safeguards in place. If you handle medical information, you likewise will be required under HIPPA to protect patient data. If your company does credit card processing at a Point of Sale system, your firm is required to participate in PCI audits. All of these regulatory and industry standards organizations help your company remain compliant, but that has not stopped companies from being hacked and it won't stop an attack on your company. If you do business within these industries, particular attention should be paid to a Cyber Insurance policy.

Employee-Owned Computers

The use of employee-owned computers in the workplace presents a series of risks. Would you need a Cyber Insurance Policy to cover the loss of data due to an employee-owned computer causing a network loss? Would a Cyber Insurance Policy cover the loss of sensitive or critical company data if it were on an employee-owned computer?

Mobile Devices

Does your company allow employees to bring their own smart phones and tablets into the workplace? Do these devices connect to company networks for email or other sensitive data? If so, your Cyber Insurance Policy may need to protect you in the event one of these devices is compromised.

Maintenance Staff

Who maintains your systems? This question can lead to a number of uncomfortable discussions. For example, if you have an in-house IT department, did you perform a background check on these individuals before hiring them? Do you maintain any type of liability insurance on them, and their ability to react correctly to an event?

In-house IT staff regularly becomes complacent with the daily tasks of their jobs. Do you take steps to keep them aware of the latest trends in IT? Are they encouraged to present new ideas and concepts to management?

If you outsource your IT support, does the company you hire also provide Professional Liability, General Liability, and E&O Insurance to cover their employees in the event an event occurs?

Consider how a Cyber Insurance policy would cover your company whether in-house staff or an outsourced staff maintains your systems.

System Age

Microsoft has either stopped support, or is close to stopping support, on a number of older computer Operating Systems such as Windows XP and Server 2003. If your company is still using old software and technology, there is a much greater risk of intrusion. Does your Cyber Insurance Policy specifically exclude coverage for unsupported systems?

Physical Security

It may seem silly to think that someone might actually break into your office and steal your server out of your data room. It should NOT seem silly that someone could break into a car with a company-owned laptop and steal it. Is the sensitive data on these computers encrypted? Does your Cyber Insurance Policy require that you encrypt your data on laptops? Does your Policy require that you be able to remotely delete data stored on a laptop? These are all questions to ask when considering a Cyber Insurance Policy.

SUMMARY OF TOP CONSIDERATIONS
FOR A CYBER INSURANCE POLICY

1. The Cloud

2. Email Issues

3. Data Type

4. Employee Owned-Computers

5. Mobile Devices

6. Maintenance Staff

7. System Age

8. Physical Security

OTHER THINGS TO CONSIDER...

The topic of Cyber Insurance may not be exciting, but the product itself should be a consideration for any company that values their data. Analyzing your risk factors helps provide a baseline for evaluating this insurance product. You should certainly work with your IT Department, or your IT Support Firm, to ensure your systems are as protected as possible. You should also look to the cyber-security industry for information about protecting yourself and maintaining a safety net.

You don't want that Thursday afternoon phone call from your bank telling you that your account has been cleared out... but if you receive it, you want to be able to get help quickly and to remediate the situation as efficiently as possible.

About Derek

As an Information Technology Professional with over thirty years of experience, Derek Davis has provided clients with application development, database design, technology solutions, and high-level management in numerous business segments, including transportation, third-party logistics (3PL), manufacturing, and telecommunications. He has worked with start-ups, regional businesses, and Fortune 100 companies.

In each engagement, Derek has gained the respect of his peers and the executive teams due to his unique problem-solving skills and insightful analysis of business needs. In several instances, his work and analysis has resulted in tens of thousands of dollars in direct savings due to improving efficiency, reducing unnecessary expenses, and streamlining work processes.

In 2001, Derek seized on the needs of the business community and started **INTELLI-NET.** As smaller company's computer systems were becoming more complex, and the importance of these systems to businesses became greater, he saw the need to help these businesses design, acquire, and implement computer systems and providing his business acumen to ensure the systems worked well and met their business needs.

As INTELLI-NET grew, and with business partner Erik Rhine, Derek realized the importance of providing advanced security to these business clients. After experiencing first-hand the effects of viruses, malware, and attacks, he realized that growing a Managed IT Business would require significant focus on security, disaster recovery, and planning.

His insights - learned over many years - are included in this book.

Derek can be contacted at:
derek@intellinet-sc.com
twitter.com/IntelliNET
linkedin.com/in/intelliNET
864-288-1114 x 100

CHAPTER 5

A BUSINESS' FIRST LINE OF DEFENSE:
DO YOU KNOW WHAT IT IS?

BY ANDREW HARPER

There are no guarantees about safety and security for any company that uses the Internet to conduct its business.

Businesses often focus on only the technology when they are taking proactive measures to avoid being hacked. This makes perfect sense, of course, but what if I were to tell you that the people that use the technology are every bit as important as the technology. It's true. **One action from a single person can open up the flood gates, regardless of how much money is spent**. What does this mean? <u>The key weakness that hackers often exploit is not technical, but human.</u>

This is the big question: *how does a business have its best chance to defend against hackers*? The answer is that businesses need to have a collaborative effort between the people they rely on to keep the business solvent and the technology that they use to make it more efficient. While there is never an "absolute guarantee" that a business will not be hacked, a strong union between humans and technology does provide businesses with a very sound chance of making the claim: "We've bested you, hackers!"

CREATING A PARTNERSHIP

*Employees and technology must be reliant on each other
in the business environment.*

Businesses, such as financial institutions for example, have a myriad of safeguards in place in order to prevent theft by embezzlement or even a robbery attempt. It makes sense and they rely on their staff to be aware of any signs that may lead to theft. *What about virtual theft?* That is an entirely different crime to tackle and one in which hackers still have an advantage, because most businesses are not up to speed with how to increase their security and create a line of defense that will better defend against a cyber attack that compromises their data. **Stopping cybercriminals is a partnership between a business' technology resources and its end users' or staffs' efforts**.

Theft of data is a huge concern for today's technology-reliant business environment. Criminals can now steal important pieces of information without ever having to step foot into a brick and mortar building. That's where the partnership between technology and staff comes into play. *You cannot rely solely on staff to perform this function, just as you cannot rely solely on technology, either*. Businesses that do this are missing the big picture and setting themselves up for disaster. And no one wishes for that—aside from cybercriminals, of course.

What are a company's most valuable assets? This isn't a trick question; its clients' data and company data. It's easy to acknowledge clients, but more difficult to acknowledge data for many business owners. They believe the myth that their data is insignificant. To that, hackers say "thank you," and IT professionals like me say: "Please stop believing that! That's a misperception that could cost you your business."

Now that you see the importance in this situation, **how do you start to change the tide and make sure that you, as a business owner, are doing everything you can to create a strong partnership between end users and technology**? You are going to proactively start enforcing six must-have layers that will ensure that everyone within your organization is "cyber alert" and better educated about the various tricks that the virtual magicians in this world play on them.

THE SIX LAYER SECURITY PLAN

By adding layers of protection, it is possible to make a network increasingly difficult to compromise.

When end users and technology work together the first line of defense in protecting a business is being created. Through this partnership, layers are put into place and each one serves as a deterrent for a hacker, decreasing their chances of being able to break into a system and reach the plethora of data it contains. Think of an onion and how there are so many layers in one that it takes a long time to get to the core of it. That is the best way to visualize creating the layers of security in a virtual world that I'm talking about. *Make it take so long to get through the layers that the hacker gives up*!

Layer #1: Education
Training end users about the very real threats that exist through technology is the first step. Once an employee is aware, they are able to take that information and make smarter choices with technology. They will recognize popular cybercriminal attempts to gain information or access—such as phishing—and make it more challenging for a weak link in the business to be found.

Layer #2: Firewall
Always use a commercial grade firewall. The devices meant for consumers are not efficient to protect a business and all its data from a skilled hacker. It's like using a squirt gun to put out a forest fire—highly ineffective.

Layer #3: Malware prevention software
Malware prevention software is also known as anti-virus software and its sole purpose is to stop any malicious virtual attempts that may hold your data to ransom, shut down your technology, or infiltrate your system to steal the data you hold on it. This includes passwords, banking information, confidential employee or client information, and anything that does not belong in the hands of people outside an organization.

Layer #4: Passwords / Passphrases
Passwords should always be kept confidential and be as strong as possible. The strongest passwords will be a phrase containing numbers

and characters, as well as upper-case and lower-case letters. Simple passwords like a name or a number sequence are highly discouraged, as they are very easy for a hacker to figure out.

Layer #5: Change passwords on a regular basis

Frequently changing passwords—at least once a quarter—and not allowing them to be reused is a smart move for businesses. It will keep any intruders who are taking an occasional "go" at your system guessing, and possibly force them to start over. They will grow sick of this after awhile. And…remember to not store passwords on the computer or in an email.

Layer #6: Computer updates

Updates to software are constantly being released to help improve it or remove a "hole" in it that has been uncovered—usually by a hacker. These patches make a huge difference between your business's technologies remaining better protected or becoming vulnerable to attempts to enter into your network through that open hole.

When business owners and managers ensure end users are trained on all aspects of the technology they are using, they are making sure that the layers of security are being enforced. There is no better partnership in deterring cybercrime than this.

CURBING EMPLOYEE RISKS TO SECURITY

An overwhelming majority of cyber attacks are due to employees, either by error or mischief.

All businesses rely on their employees to be successful, but unfortunately, employees are also one of the biggest risks to a company. **That is why it is continuously emphasized in the IT community that security awareness is an essential part of the company culture**. You must operate a security-conscious company. Here are three effective ideas to promote the security conscious company that your business does need:

1. Periodic education sessions

Having end user responsibilities be a part of new employee orientation should be a must. And when it comes to existing employees, having quarterly education classes or lunch-and-learn sessions is an excellent way to keep technology users "in the know" of new threats or provide reminders of why certain security measures are in place regarding:

- Email use, including phishing, compromised links, and other suspicious looking emails. *Do employees know that they should always check with an IT specialist before opening anything suspect?*

- Password privacy and changes.

- Giving information to unknown parties, whether via telephone or email.

- New threats and viruses that are out there.

2. Awareness literature

We all learn and retain information in different ways. Having face-to-face education sessions is great and it becomes even better when it's followed up with newsletters or visually explanatory posters throughout the business's premises. When employees get busy, they focus on getting their work done and sometimes that can mean that cyber security measures are put aside. The more chances a business has to engrain security in its employees' minds, the better the chances of them being remembered becomes.

3. Acceptable Use Policy

Have a thorough and thoughtfully laid out Acceptable Use Policy that is signed by all employees and placed in their files. This will let them know what is acceptable, or not, with business technology. This document is strongest when it is drafted with the assistance of an IT professional, because of their vast experiences in cleaning up after cybercriminals.

In addition to the three ideas mentioned above, there are two other actions that management and business owners should take in order to reduce a business's risk of having their data compromised by an end user. They are:

- **Have role-based access**: Limit employee access to data and information, and never provide any one employee access to all data systems. There is no reason for an employee to have access to more systems and data than what is required for them to perform their job.

- **Remove administrative rights on computers:** Only individuals that work with IT should install software onto a business system. This is probably the single most overlooked strategy for cyber security out there, which makes it a huge concern.

Will having all the technology and information ingrained in employees' minds make a business 100% safe from a hacker? No, it won't, but it will absolutely make your chances of avoiding a cyber attack better than they would be without it. Yes, there will be times when employees click on a link they shouldn't, install malware or activate a virus or take some other action, whether accidental or intentional, that puts a company at risk. This is where the technology portion will enter into the partnership and help a business out!

ACTIVATE YOUR TECHNOLOGY DEFENSES

*Technology prevents employee errors from
impeding business security.*

Having the proper technical safeguards in place will help a business ensure that its technology is looking out for its employees just as much as its employees are mindful of the technology they rely on. Again, it's a partnership in which they have each other's back.

There are three ways to activate your technology defenses and keep them standing strong. Not doing these things **is not an option** if you are truly vested in the security of your business.

1. **Patch and update**.

   If your business is not large enough to have your own full time IT department, you had best have a relationship with a professional IT company that can make sure all updates and patches are performed as soon as they are out. There should never be an "I'll get around to it" philosophy—that is unacceptable if you're committed to security.

2. **Use business class security, not consumer grade equipment**.

   The best IT companies will insist on providing the equipment you need, including: firewalls, email filtering, web filtering, and intrusion prevention technology. Why? Because they know that they are installing a clean device or system—one that has had no chance of having malware put on it before installation.

- **Secure your website**.

  Not all website hosting is created equal. When you are looking to secure your website, make sure that you choose a host that has:

  - An emphasis on security
  - Performs auto-updates
  - Has malware protection
  - Performs automatic backups with "restore" ability

THE BIG PICTURE

Humans are the best first line of defense.

Business and technology go hand in hand. Remember that cyber safety and humans are intricately linked. The technology is created by humans, implemented by humans, and used by humans. There is no removing the human factor from technology. That is why a partnership between educating humans and having strong technical parameters is the ideal collaborative effort for a business that wants to succeed.

About Andrew

Andrew Harper is co-founder and CEO of **Gaeltek, LLC**, an award-winning managed services provider located in Manassas, Virginia. Gaeltek monitors, manages and maintains the network technology of small to mid-sized businesses in the Washington, DC area and is dedicated to providing its clients with cost-effective, state-of-the-art solutions.

Prior to co-founding Gaeltek in 2004, Andrew served over sixteen years as a Weapons Engineer Officer in the Royal Navy. His service included assignments on Her Majesty's submarines and surface vessels as well as teaching positions, including at Royal Naval Strategic Systems School. His final Navy posting was to the British Embassy in Washington, DC as the Technical Liaison Officer and Head of the Technical Department for Strategic Programs, Royal Navy. Upon completion of his service in the Royal Navy, he joined the British Embassy in Washington, DC as the IT systems manager for the British Council, the international educational and cultural arm of the British Embassy.

Andrew and the Gaeltek team have earned many honors including the Prince William Chamber of Commerce 2013 Business of the Year; Next-Gen 250, one of the 250 IT companies nationwide honored for utilizing a smart and different approach to IT services and integration while providing exceptional service; and one of the 40 most innovative IT companies in the United States and Canada by industry publication CRN for ingenuity and success in developing unique ways to utilize technology to help customers solve business problems. Andrew has been named to the MSP Mentor 250 list, an industry honor that identifies the top 250 leading IT executives, entrepreneurs, and experts worldwide as well as the SMB Nation Top 150 list of the most influential people in the small to mid-sized IT channel globally. Andrew has been featured in such publications as *Channel Pro magazine, Prince William Living, Be Better at Business*, and on Executive Leaders Radio, the number one business weekly radio show in the Mid-Atlantic region of the United States. He also has served on numerous industry advisory boards and the ITT Tech Program Advisory Committee.

Mr. Harper is co-author of *Business IT 101: A Business Owner's Guide for Finding Hassle-Free Computer Support* (2011). He holds a Bachelor of Engineering (Honours) from the Royal Naval Engineering College.

CHAPTER 6

CONTENT FILTERING AND EMPLOYEE MONITORING

BY BRYAN HORNUNG

I have encountered all kinds of excuses as to why a business owner does not want to know what his or her employees are doing with their company assets and company data. Usually this attitude changes the minute something awful happens that probably could have been prevented. Some business owners choose to be oblivious to what their employees are doing and delude themselves into believing their employees only do work-related tasks every minute of the workday.

I've had business owners try to tell me that unmonitored Internet is a benefit they give their employees, like paid time off. If you are one of those business owners, let me caution you that it is not a benefit and if you have convinced yourself you have assembled the most honest and productive staff who would never use the Internet for personal business, then I can help you no further. Don't waste your time reading further because you have some sort of artificial intelligence, not human beings working for you. Content filtering and employee monitoring is an important subject that you may not want to think about, right? To ignore is bliss? Is the total loss of data, or giving criminals access to confidential information about your company and your clients blissful?

If you really value your clients and your company, you cannot take a *laissez-faire* approach to content filtering and employee monitoring.

CONTENT FILTERING IN TODAY'S WORLD WIDE WEB

It is one of those days you are feeling great and when nothing can go wrong. You just walked back into the office from an important sales meeting in which you landed a new client. You are pumped and nothing can bring you down. Well . . . almost nothing.

The first person you see is your marketing director and you eagerly tell her the news that the potential client you had been targeting for months has just signed a huge contract. Oddly, she doesn't seem as pleased as you would expect. She pauses for a moment before she begins speaking and suddenly the world around you slows down to a crawl. You learn all the marketing files created over the last 10 years are not opening. None of them. Yep, you just slipped off a cliff and are headed down.

A few hours later, your IT support calls and informs you that you've been infected with Cryptowall 3.0, the latest flavor of ransomware that encrypts any file it can get its hands on and then prompts you to pay upwards of $1,000 or more to get your files back. You just landed face first into the dirt. The first question you ask, "How could this happen?"

A few years ago, Web filtering was considered rather easy. IT people could block Internet traffic based on categories, like gambling and perhaps social media. They could tap some keys and whamo! . . .the sites were blocked. By no means was this method perfect for blocking websites, but until several years ago, it is all many corporations had to combat employee Web surfing.

The landscape of today's Internet is much different, and fraught with cybercriminals who are constantly hacking into legitimate Web sites that would not normally end up on your blocked list. Everyday, cyber criminals are targeting thousands of small to medium-sized businesses that utilize Content Management Systems such as *WordPress* and *Drupal*.

80% of all web malware is now hosted on legitimate web sites that have been compromised. The hackers typically exploit websites that are left unpatched and are not regularly updated. Because many companies fail to do this, they make it easy for cyber criminals to exploit your company by launching an infection when the compromised website is visited by one of your unsuspecting employees. This type of infection can happen any time, to any website, including your vendors, partners, and clients. A website that your employees visit regularly may have been safe yesterday but could become infected today. The worst part is that it could infect your system without anyone's knowledge until the damage is done.

There are millions of Web sites that are infected with malicious code that can wreak havoc on your network, and there are tens of thousands more discovered everyday, so utilizing Web filtering that only blocks questionable content will not keep your company safe from web-borne threats.

THE SOLUTION

So what can you do to keep your employees safe but also allow them to stay productive while doing their jobs?

1. Implement technology that performs "Deep scanning" of web traffic as it is accessed. This is very similar to your antivirus having active protection. The Website traffic is scanned before it is presented to your employees on their computer and in their Web browser. If the Website they visit has something malicious within the site, it will be blocked and crisis potentially averted.

2. Keep your computer's operating systems and the software installed on it up-to-date. This will protect against exploits that only target unpatched or unprotected systems. Limit the number of Internet browsers to a standard set and enforce their use as a policy.

3. Make sure remote users are also protected by following the above, as well as enforcing the use of a VPN to access the Internet or install software that controls web policy enforcement and content scanning from the cloud.

COMPUTER & INTERNET-USAGE MONITORING

The Internet has changed the way we live, how we communicate, and changed the way we conduct business. For most businesses, the Internet has become one of the most integral parts of the company. However, the effective and efficient use of the Internet by employees is not guaranteed.

In a recent survey conducted by Salary.com, 64% of employees surveyed visit non-work related websites every day. Roughly 40% of those surveyed spend one hour or less per week while 29% spend two hours and 21% waste five hours per week – while only 3% said they waste 10 hours or more doing unrelated activities.

Most employees spend their time socializing on Facebook and LinkedIn. Other popular destinations include Yahoo, Google+, Twitter, and Pinterest. The rest of the time, employees spend their time shopping at sites like Amazon.com. According to the Salary.com survey, the younger workers between the ages of 18 and 35 tend to be the biggest group of recreational Web surfers. This simply means that as the younger generation takes over the workforce this type of behavior will become more prevalent in your business.

There are two common arguments for monitoring your employee's computer and Internet usage. The most common concerns are productivity and protecting yourself from litigation. Although on the surface you can use the same tools to protect yourself from litigation and combat unproductive web surfing by employees, I would highly discourage you from doing so.

Instead, help make your workforce more productive by other means such as allowing short breaks throughout the day to help them achieve higher levels of productivity. As it relates to inappropriate website browsing, you should implement Web filtering to block or track websites that are offensive or not related to work activity, such as pornography or hate speech. If you suspect excessive wasted time by a particular employee, you can have your IT team install activity monitoring software such as SpectorSoft to monitor the activity on that particular computer. This type of software is far more aggressive and can provide much more data and a clear picture of what that employee is doing so you can take corrective action. You do not want to use this software on every computer in your network because it requires a great amount of resources which can slow the system down.

E-MAIL & FILE MONITORING

In my experience as an IT consultant, one thing that I have noticed is that business owners do not seem to think that what their employees are doing is taking company data and transmitting it outside of your organization.

Recently I received a phone call from a client who asked us to check the email of a former employee to see if he was emailing himself or stealing any company data. This isn't the first time I received a call like this. Unfortunately, this client, like most, did not implement any monitoring of employees within its network. Up until that point, the client had informed us that they did not want to operate "that kind of business" and they felt that their employees should have the freedom to go on any Website. Our client, the business owner, had no interest in knowing which sites his employees visited, how often, and when. That was until the day they called us, concerned that a member of his management team had left the company for a competitor. At that point, while the owner's concern was merited, it was way too late.

Trying to figure out if an employee is transmitting data outside of the organization is not as simple as going into a sent items folder and seeing if there is anything in there that would prove to the employer that the employee was stealing company-owned data. There are too many ways in today's world that an employee can take data from your network send it anywhere in the world with the click of a mouse. Services like DropBox, Google Drive, Gmail, and Skype make it far too easy for employees to steal data from your organization. Controlling access to particular web sites is only half the battle. You also need to protect your data from being stolen via removable devices such as USB drives. Technically savvy people know how to access their favorite websites through smart phones and proxy websites. Therefore, it is good business practice for companies to create, publish, and enforce policies on personal Internet use as well as the use of removable devices.

An additional way to protect your data is by implementing file encryption on your company data. Data breaches at Target, The Home Depot, and JPMorgan Chase could have been prevented if certain data was not so easily accessible to hackers. As such, these companies suffered some pretty bad PR and had to spend millions recovering and protecting the damaged caused by the breach. Smaller companies can suffer the same fate, only with less fanfare.

File encryption ensures that files can only be opened on computers that are authorized to open it. This usually requires your IT administrator to install encryption software on each device. Without the presence of this software a user cannot open the file. The tradeoff can be in performance and can cause files to open slower than if they were unencrypted. Work with your IT consultant and determine which files require encryption and which ones do not to achieve the optimal balance between performance and security.

SOCIAL MEDIA MONITORING

Proprietary data theft, conflicts of interests, or even lack of productivity are not the only thing you need to worry about as the owner of a business. Your corporate reputation could be on the line when it comes to what employees post on their social media accounts.

Think about it. Every single employee in your company has the ability to publish to social media. It doesn't matter whether they're officially speaking on behalf of the company. Like it or not, your employees are representing your company and it doesn't matter if they do it from home or from the office. What an employee does on social media can harm the reputation you have worked so hard to build in an instant, especially on business-oriented sites like LinkedIn.

Recently, an employee at Brookfield Zoo in Chicago posted an Instagram of herself with a racist remark about the clients she was serving. She identified herself as an employee of the Zoo and posted it while she was working. She was promptly fired for the remark, but her remark became viral, worldwide in minutes. The damage was done to the reputation of the zoo, and the controversy made national news.

Employees are not going to be too thrilled if you require them to have their social media accounts monitored. Many states are even implementing laws that limit an employer's ability to access social media accounts. While this area of monitoring is new and ever-changing, the best course is to have a clear social media policy distributed to employees. As an employer, you need to have a clear and detailed outline of what is and what is not expected and acceptable on company-owned devices as well as social media. This includes posting on social media while they are on the clock, even if it is on their personal device, and that they should not be directly representing your company and brand unless given permission to do so.

Filtering content and monitoring employees requires a delicate balance between your corporate interests and the rights of your employees.

Your company's policies and procedures must be optimized for both corporate reputation and employee rights and you cannot combat one at the expense of the other. Establishing and enforcing policies should improve productivity, increase the security of company information, boost the security of company technical assets, and reduce the issues associated with sexual harassment or poor job performance.

It is important to understand that a complete content filtering and employee monitoring solution is one that involves both software and written policy. Working with an IT professional and legal counsel to have both of these in place in your business will not only prevent abuse of company-owned devices, but they can also help you effectively determine if employee theft or conflicts of interest are taking place right under your nose. You can then take immediate action before suffering huge financial loss to your business.

About Bryan

Bryan Hornung is CEO/President of **Xact IT Solutions**. Bryan has consulted with and helped hundreds of small and medium-sized businesses demystify technology since starting Xact IT Solutions in 2004.

Bryan's career began in 1999 as an IT consultant for the NAVSEA's Naval Surface Warfare Center, Carderock Division (NSWCCD) where he was instrumental in implementing Web-based technologies to help coordinate projects between engineers and the U.S. Navy fleet. His work earned him the opportunity to work closely with Navy Captains and their civil counterparts, arming him with the confidence and knowledge to start his own business.

Bryan's excellent reputation at NAVSEA paved the way for many opportunities and he began moonlighting as an IT consultant for a few small business beginning in 2002 – while employed as a DOD contractor. A couple years later, he grew his moonlighting opportunities into a full-time consulting business and started Xact IT Solutions Inc.

As CEO, Bryan oversees the daily operations of the company while also consulting with C-Level clients as a virtual CIO for their businesses. Bryan's focus is always on making sure his clients receive the best service possible and constantly helping them improve efficiency and their profitability through the use of technology and the right services. This dedication to his clients has been recognized throughout the years winning multiple awards year after year for their outstanding customer service and brand awareness.

CHAPTER 7

DEVELOPING AN ACCEPTABLE USE POLICY

BY DAVID ROSS

You might be asking yourself, what is a chapter about Acceptable Use Policies (AUP) doing in a book about cyber security attacks? Good question. . . An AUP can be your most effective defense against such attacks. It can be your first line of defense. This is because, without humans in the loop, viruses, malware, Trojans, and other computer threats have no purpose. The more knowledgeable the human user, the less likely they will fall prey to such attacks. If no one clicks on the link or goes to the website, the virus cannot get in.

The best protection you can have against cyber attacks is an educated user. An AUP does exactly that by outlining policies for the use of all technological and organizational resources, including personal resources used on the organization's network or within the organization's facilities. An AUP then becomes the first line of defense against cyber threats and even against threats to the organization's physical assets.

For example, consider a business that accepts credit cards for payment. If an employee in this organization, who doesn't know better, records credit card data in a Microsoft Word document, which can easily get lost or stolen, then the employee has put the organization at great financial risk. Simply recording credit card data in that manner is a violation of the Payment Card Information Data Security Standards (PCI DSS) and could put the organization out of business for violation. An AUP policy, which all employees would have to sign, could easily prevent this type of situation. At the very least, having an AUP policy that outlines the usage of credit card data would show due diligence on the part of the organization and help provide protection in the case of misuse or loss of data.

The overall purpose of the AUP is to provide reasonable accessibility to the resources required to operate the organization, without compromising the integrity of the organization. A well-developed and comprehensive AUP can save your organization. An effective AUP should include policies for everything you need protected and outline compliance issues, such as enforcement, and note how to deal with violations.

WHAT SHOULD BE INCLUDED IN AN AUP?

An AUP policy provides the same thing for any organization, but the requirements and details that should be included come in an endless variety. Each industry has specific needs, creating a need for different AUP policies. Developing an AUP requires you to think through each piece of your organization:

- The resources involved.
- How critical each resource is.
- How the use or misuse of a resource can affect operations, morale, reputation, and profitability.

An active, observed AUP is one that is talked about, monitored, and has everyone's buy in. This will bring awareness to behaviors with potential consequences to the organization, both internal and external threats. This awareness gets people talking, further spreading awareness and sharing how to spot things that look out of place, so no one touches it, thus preempting the problem.

An organization can have a need for multiple AUPs due to differing requirements for differing users. A separate AUP is needed for each organizational interface, such as employees, partners, vendors, clients, etc., because each has a different relationship with your organization and thus touches different resources than the other. Employees have the greatest connection to the organization and, consequently, have the largest impact on compliance with the AUP.

Although different organizations need many different things from an AUP, all types of organizations need many of the same things. Every organization needs policies outlining the proper usage of resources such as the following:

- The Internet, including websites, social media, streaming video, music, etc.
  - Employment contracts should contain clauses pertaining to the appropriate use of social media and behavior on social media sites for business purposes and personal use on company time. It should spell out unacceptable online behavior and the consequences to employment if not observed.
- Communication devices such as office phones, faxes, cell phones, and tablets.
- Computer resources such as desktop computers, laptops, printers, and network.
- Physical resources, such as company vehicles, supplies, tools, and facilities.
- Personal assets such as employee-owned laptops, tablets, and mobile phones.

Use of personal assets, such as laptops, tablets, and mobile phones, should be included in an AUP for many reasons. Personal assets can pose a real threat to organizational resources. This is because personal assets are not managed by the organization's support department, and consequently, have not been configured to protect the organization's assets. For example, viruses can spread from the personal asset into the organization's network if the device has been infected. While it's difficult to control what people do with their personal assets, when used in conjunction with the organization's resources, such as the Internet, it is important they understand that inappropriate behavior is in violation of the AUP.

Special attention should be paid to hosted/cloud resources. Today, more and more applications are hosted in the cloud, such as scheduling, email, or training tools. Most businesses today operate in a hybrid environment, with a mix of in-house and hosted or cloud resources.

WHAT ABOUT REGULATIONS AND INDUSTRY SPECIFIC NEEDS?

Organizations subject to regulation, such as healthcare, government, finance, banking, and investment need to pay special attention to how regulatory considerations can affect an AUP. Depending on the regulations, there are specific things that must be included to ensure compliance. Here are some specific regulatory agencies and regulations for these industries to keep in mind when developing AUPs:

- Finance –Financial Industry Regulation Authority (FINRA)
- Health – Healthcare Insurance Portability and Accountability (HIPAA)
- Government – Federal Information Security Management Act (FISMA)
- Banking/Investment – Security and Exchanges Commissions (SEC)

Each industry may also have recommended inclusions to take into account when developing an AUP. Some examples are listed below:

- Education – Each level of education will have different needs. K-12 will need to protect the child's privacy, keeping them from accessing undesirable web sites while not restricting reasonable research activities. One of the difficulties of developing a K-12 AUP will be educating students on what unacceptable behavior includes. How detailed the education on unacceptable behavior should be will depend on the age group. It would be inappropriate to teach an eight-year-old how to identify an adult site. One study showed that when developing an AUP for educational institutions, getting the community involved early in the development process can have a real impact on the AUP's acceptance and consequent adherence.
- E-commerce – Costumer data, credit card information, inventory management.
- Health Care – Handling of physical records, electronic records, medical instruments, sanitary practices.
- Real Estate – Treatment of the client's property and their privacy.

HOW DOES AN AUP PROTECT?

What does an AUP have to do with cyber security? Simply put, an AUP is the first line of defense against a cyber-attack. 90% of viruses come through computers or infected websites. The vast majority of viruses enter a network because someone fell victim to phishing and clicked on something they shouldn't have. AUPs help defend against these avenues of intrusion by guiding all avenues of entry and all computer resources.

An AUP protects an organization by outlining the use of personal assets and corporate resources by employees, contractors, and clients. For example, many businesses now offer customers free Wi-Fi while in their facility. A smart business will create an AUP which customers must read and agree to before accessing the internet. This protects a business by clearly stating the appropriate uses of the business's Wi-Fi and consequences of violation. It can stop someone on the guest network from sending spam from the business location, which can lead to the business's emails getting blacklisted and requiring timely and expensive work from IT to remove the business from blacklists.

Part of the importance of an AUP is the act of due diligence. By creating AUPs for employees, contractors, customers, and others, a business is doing its part to protect its assets. By failing to put AUPs in place you are basically leaving it up to someone else to look out for you, which at some point will not suffice and render you negligent.

WHOSE RESPONSIBILITY IS IT?

Who develops the AUP, monitors, and enforces it? There is no one-size-fits-all answer. It varies by organization type and culture. A school AUP would need to be created, at least partly, with parents' input. Other times the best person to create an AUP is the corporate lawyer or HR. Regardless of organization type and culture, IT needs to be a front runner in development of the AUP because they can advise what is and isn't possible, the cost for each option, and how implementation and detection would work.

No matter who develops the AUP, its success is largely dependent on the end users, so it is important to get their buy-in from the beginning. The more buy-in of the contents, the greater the probability that the AUP will be followed. The individual managing the AUP development project needs to be a constant voice of reason. The broader the community of the development team, the more of a balanced voice they must be. The goal is a "usable and active" AUP.

After an AUP is developed and put into place, monitoring and enforcement become important aspects of managing the AUP. In most cases, IT will work to monitor compliance and detect violations, while HR will serve as enforcement and manage consequences for violations.

Compliance to the AUP can be measured in a variety of ways. Utilizing a proxy server to track all websites visited is often a component of compliance. Developing Key Performance Indicators to measure compliance to an AUP is another important component. In order to improve use of and compliance with an AUP, you need to measure usage and determine metrics for how well you are doing overall toward the goal of 100% compliance. The ultimate goal is always zero violations.

Unfortunately, achieving zero violations is not always realistic. When violations occur, there should be consequences, which should be outlined in the AUP. The effectiveness of the AUP is directly related to compliance efforts. If consequences for violations are not enforced, then no one will care that there is an AUP, and no one will follow it. The AUP needs to be clear, fair, and can NEVER be compromised. If it is compromised, then no one will care about the AUP, and buy-in will be ruined.

It is typically the role of HR to enforce consequences for violating the AUP. They will decide how severe a violation is and what action to take. Consequences could be as severe as termination. Lesser consequences could be a simple discussion with a supervisor or HR. For some industries, consequences for violating an AUP could include a fine. An employee found to be in violation of the AUP could also face less tangible consequences, such as a negative effect on their overall reputation as an employee and on opportunities for a raise or promotion.

WHEN SHOULD AN AUP BE CREATED?

The simple answer is immediately. If your organization does not currently have an AUP in place, one should be created immediately. If you only have one in place for employees, you should consider developing one for customers and contractors. Chances are, if you think you need an additional AUP, you probably do.

If you already have an AUP, or multiple AUPs, in place, good work. That is a big step out of the way, but does not mean the work is over. It is now time to focus on educating your employees, customers, etc. about the AUP, and on monitoring compliance to the AUP and enforcing consequences for violations. Also, if it has been awhile since your AUP was developed, you should review it to make sure it is current. For example, if you have an older AUP, does it address usage of applications in the Cloud?

Whatever stage you are in, remember IT plays a vital role in the creation, implementation, and monitoring of an AUP. Don't forget, the biggest indicator of a successful AUP is user buy-in; without it compliance will suffer. Remember a successful AUP is ultimately your first line of defense against a cyber attack.

About David

David Ross is the President and CEO of **Continuous Business Inc.** As his company's name implies, David's mission is to provide IT solutions for a business that is specific to its unique situation, and that can be accomplished without disruption to its daily operations and without incurring high costs. He incorporates a business' existing software, networks, and business processes within his design and engineering of enterprise grade systems.

David has over 29 years of experience in the information technology industry in many diverse business settings. Under his management of the company, Continuous Business integrated a heavy equipment manufacturer's new web-based loan origination application with multiple existing independent applications into one viable system. Continuous Business also extended the life of a manufacturing company's aging infrastructure by designing a fault-tolerant, core-computing platform that upgraded the computing and data resources for the company's diverse business applications. This saved significant capital investments for his client and contributed directly to bottom line profits.

David's efforts increased Continuous Business's sales from $385k to over $3.7 m. Prior to this venture, David worked for Syscom Inc. as Director of Research and Development. It was his responsibility to develop multiple software innovations that allowed the company's document management system to integrate with diverse third party applications. Previously, David worked for CSX Technology as the manager of their Distributed Systems Group. Here he directed a team of twenty engineers in such leading-edge projects as: radio-based voice recognition dispatch; wireless handheld computer-based train yard repair system; and a wireless tablet-based rail car repair shop management system.

CHAPTER 8

THE NEW FACE OF BUSINESS: "BRING YOUR OWN DEVICE"

BY JAMES MARTIN

In today's business culture, speed and efficiency,
in a cost-effective manner, are expected.

Not every industry is new to having employees bring their own tools to work. My brother is an airline mechanic, and for years he's purchased his own tools to use instead of using the ones that his employer had available at the shop. Why would someone do this? The logic is simple: they feel that the tool they have is better than the one in the shop, and they know exactly how it works by its feel. This concept makes sense when we think of it that way, and with the new face of business, it has become a part of many businesses.

Our modern business environment is highly reliant on technology, and expects that its employees can effectively use it in their positions to contribute to the business' success. What happens when we take the tools concept and apply it to our modern business environment? The result is that we are entering into one of the most popular and interesting newer dynamics in business—*the Bring Your Own Device (BYOD)*

71

culture. This is exciting and when properly implemented, many positive outcomes result from this movement, including:

- **Increased productivity**: Having equipment that you are comfortable using, in a variety of locations, creates a faster, high quality work pace.

- **Devices are taken better care of**: When an employee owns a device, even one used for work, they will treat it better.

- **Cost efficiency**: The costs of equipment purchasing, updating, and maintaining goes down significantly for businesses with BYOD policies.

- **Employee satisfaction**: Just like the mechanical tools, employees like it when they are comfortable on the technology they are using. After all, it was their choice!

- **Single device solutions**: Being able to use one device for many needs is ideal for most technology users. It keeps things simpler because it eliminates having to keep track of multiple devices.

Business owners want these types of benefits in their work environment. They can't go it alone, though. They rely on their IT professionals to deal with all the important details that come along with having a BYOD work culture. That is what I want to walk you through today. **How do we go from idea to inception**? What type of collaboration is required between business owners, CEOs, and IT professionals?

THE IMPORTANCE OF THE IT ROLE IN BYOD

One of the most noticeable things about BYOD is the evaluation process it goes through to even become accepted into the workplace from management and decision makers. Most employees are advocates of it, liking the flexibility and opportunity it offers them. They just need to hear about it and they think, *that would be fantastic*! In turn, they talk to management about all the reasons why it is a good idea—sighting the obvious factors of cost and productivity. Finally, management agrees to investigate its potential a bit further. It is easy to see why they should seriously consider it, because information can be found in many places. Some of the more notable statistics that show why proactive management does give serious consideration to BYOD include:

- 19% of businesses believe that BYOD has increased employee satisfaction (CCMI).

- By 2017, it is expected that only 50% of firms will provide their employees the technology they need to perform their jobs (Gartner).

- Already, 90% of workers in the US use their smartphones for business purposes (CISCO).

- Approximately 59% of all IT decision makers have found that it is a competitive disadvantage to not embrace BYOD (Dell).

- An estimated 40% of employees log onto their email from home when they are sick (Click Software).

- Nearly 82% of companies allow their employees to use personal devices in the office (Intel).

These statistics show that whether a business is BYOD resistant or not, it is an inevitable topic that will need to be addressed, if it has not been already. Admittedly, it may be overwhelming and something that decision makers are hesitant to place on their to-do list. It seems like a huge feat—and it is. It requires a sound collaborative effort with experts in this area. This is why IT professionals, whether contracted or on-site, are pivotal in partnering with businesses for BYOD implementation. Without that collaboration, the possibility of this new face of business being set up improperly increases. <u>And as a business, do you want to be a leader, a follower, or fall off the radar and become yesterday's news?</u> Just as IT professionals know, business innovators know that it is all about understanding the risks and building a plan from there to lead to success.

KNOW YOUR RISKS

Risk comes from not knowing what you're doing.

~ Warren Buffet

Is there risk with BYOD? Absolutely; however, knowing what the risks are is the first step in guiding IT and management in the process of creating a plan that will help you do what you want to, which is create workplace productivity and efficiency in a cost-effective manner.

There are five risk areas that a business needs to consider when they are making the determination of whether BYOD is a smart choice for them. You have to weigh and figure out which ones are most significant to the business's livelihood, since the primary goal is to streamline and make processes better, not create problems.

1. Regulatory requirements

Most businesses have some sort of regulatory requirements. Those who work in the healthcare or finance industry definitely know this! Make sure that you are following all the safeguards and requirements that are in place for securing data in an appropriate manner that protects customers and the business, too. Not doing this leads to potential fines and a loss of reputation that is often impossible to fully recover from.

2. Applications

With a BYOD device, employees may have personal applications on their device, which means that you have to find ways to ensure that those non-business apps are not bringing in malware of any sort to the device, or compromising the work-related data the device has access to.

3. Data loss and recovery

Systems need to be implemented and followed to ensure that proprietary and sensitive data is protected. A plan to recover or wipe work-related data from a user's device due to theft or employee job change is also necessary.

4. Privacy

This is a huge concern for business to address with BYOD, because IT will most likely need to monitor and track a device's location, which does tie into the user's location—even during non-work hours.

5. Labor laws

Depending on local and state laws regarding labor, BYOD can present complications for ensuring that employees don't work over a certain number of hours. This needs to be tracked and is more challenging to do, particularly if an employee can work off site (which is one of the greatest benefits of BYOD in the workplace).

Once you have risk assessment out of the way, the IT department and relevant business staff can move on to the next step, which is creating a BYOD Policy for the organization.

THREE AREAS TO ADDRESS IN A BYOD POLICY

You can take the "unknowns" out of BYOD
by having a clearly defined policy.

The general term "BYOD Policy" is quite vague and it is up to each business to determine what the acceptable parameters of their own policy will be. *There are some businesses that will decide that no BYOD is the only way to circumvent risk.* This is often the case with businesses that handle data that is extremely sensitive, such as US government data and information, or a business that has employees that travel overseas a great deal.

More often than not, businesses are going to see the value of having some type of BYOD Policy and as this work concept grows in popularity, they are going to find a way to incorporate it into their culture. It does make sense to do this if the proper procedures are put in place. **Once a decision is made to consider BYOD, it is time to intensely scrutinize what is important to the organization**. From there, the determination of how to implement the policies can take place. The main areas that businesses and their IT decision makers should focus on are:

The culture of how the devices are to be used must be established and nurtured. This means that all employees who are a part of BYOD will understand how the process works, the benefits of it, and the expectations that their employers have of them.

The written policy must be constructed in a way that is understood by all employees. This is for everyone's protection, and a sound step to take in order to avoid the "I didn't understand that" or "You knew that" moments that can happen when things go awry. It is strongly advised that a business have an employee signed copy of the BYOD Policy in each of their files as an extra safeguard. Another recommendation is to work closely with a qualified, highly knowledgeable BYOD IT firm to create the policies, as these professionals have experience helping other clients successfully implement these types of plans.

Policies and understanding are of little value if the proper technology is not put into place. Transitioning into a BYOD culture should be—and can be—a smooth, user friendly process if it's done right. There is no better way to achieve this than working closely with your IT firm and allowing them to put the proper technologies in place. But equally important—they also need to be tested to ensure they are doing what is expected on a consistent basis—monthly is the best.

PUTTING THE TECHNOLOGY IN PLACE

Technology brings security and peace of mind to BYOD.

Your business has embraced BYOD and all the positive aspects that come with it. Your policy is in place and you know who your users are. Now it's time for the IT professionals to put the technology in place that makes this bold new way of business work the way it should. There are two steps left for IT to perform and then you'll be a part of the most in-demand, currently-existing business trend.

Step Number One: *Determine what mobility management platform will be used.*

Through comparison and analysis with IT, a platform determination can be made that will:

- Be compatible with a business' current technology and security needs.

- Support any plans for future requirements or necessary adjustments.

- Will be able to support the user demands and emerging technologies.

- Provide strong network access controls.

Step Number Two: *Have a user-owned devices plan that's supported by the business.*

The implementation of BYOD does not mean that a business no longer needs to provide device support for employees. Making this mistake will lead to reduced job satisfaction and increase the potential for a breach. IT will help you set up the proper support by addressing and overseeing:

- The protocols that help everyone properly manage each phase of the BYOD implementation.

- How devices are set up within guidelines and decommissioned after they are replaced or an employee is no longer with the organization.

- Employee management of the devices to reduce risk of vulnerabilities, while improving on safe practices.

I often get asked how long it will take to implement a BYOD policy, and the best answer truly is, "It depends." We have to factor in many details, such as: **current technology, number of employees, the objectives of the project, and the transition from business-provided technology to business-provided support for employee-provided technology**. Most often, a business will start with a smaller test group to better learn how the transition will impact the business and then slowly incorporate everyone in. The one thing that everyone agrees on is that rushing the process is not beneficial. It's just like the old adage most of us likely heard growing up. *It's better to do something right the first time than have to go back and do it again.*

ARE YOU READY FOR THE NEW FACE OF BUSINESS?

Fewer trends are more exciting than BYOD for business.
It truly is the way of the future.

Everything that seems like it might be startling about BYOD can be alleviated with the proper partnerships between IT decision makers and the qualified IT professionals that understand this latest trend. Through my business, FIT Management, I have worked with many companies, helping them make smoother transitions into the BYOD culture.

Through collaborative efforts, IT and management can merge together to create an exciting business environment where things are "clicking." *Employees are happier and more productive, which leads to better bottom lines for the business itself.* Knowing that BYOD is here to stay for the foreseeable future, isn't it better to start embracing it and preparing for it? I say, "Yes." **It's easy for me to be excited about this, because I know how to reduce the unknowns and help deliver the outcomes that make everyone love this new face of business**. As for you—when you make educated moves through the process of BYOD for your business, your excitement will also grow.

About James

Like yourself, Information Technology expert and business entrepreneur James Martin has seen technology change numerous ways over the years.

From the days of the single desktop computer to today's smart phones and the Internet of things (IoT). James is always looking at what is next and how it will advance his customer's business.

With over two decades of executive training, infrastructure planning, and most importantly, real life, in-the-trenches business experience, his view is radically different. James appreciates, and shares with clients, that technology leadership is for everyone's business, everyday. It's a major way to improve your business.

Technology is an ever-moving process that must be evaluated to make sure the technology you are using is improving your business without leaving you vulnerable.

CHAPTER 9

WHY SMALL BUSINESSES ARE A CYBERCRIMINAL'S FAVORITE TARGET

BY KRIS FENTON

Your business could be considered the best of the best for many years, but all it takes is one cyber breach to change the perception in an instant. Then what happens? You go from the "professional" business to the "reckless" business from that single event. Sure, many of the businesses that have been attacked by a cybercriminal that we hear about are the large corporations—the SONYs and TARGETs—but those mega corporations have woken up! They are growing smarter and investing their resources into stopping these illegal invaders from getting into their technology. <u>So, where are cybercriminals turning</u>? **They have a new target, small to midsize businesses**. According to The Small Business and Entrepreneurship Council, businesses with less than 500 employees make up 99.9% of all business in the US. And of that percentage, businesses with less than 20 employees is 98% of that. That's a big selection of businesses for these Internet criminals to choose from!

Smaller businesses are in cybercriminals' crosshairs more often because criminals know they are vulnerable and that the mentality of many smaller businesses is that they are not significant, which makes them more vulnerable—actually, it makes them the "ideal" target. So, even if a criminal can't steal a billion dollars from a bank account (which has happened) or obtain trade secrets from a foreign nation, cybercriminals can find a rewarding pay-out on a smaller level. *And worse—they can often go undetected for quite some time—up to 200 plus days.*

WHAT ARE YOUR ODDS?

The statistics are startling.

Misperceptions by small and medium-sized business owners exist all the time when it comes to what they "perceive" the value of their data to be. They underestimate it, which means that they are not taking this "high threat" seriously. According to the National Cyber Security Alliance, **1-in-5 small businesses fall victim to cybercrime each year**. And, nearly 60% will be out of business within 6 months. The smaller the business, the closer it creeps up to 100%. The natural question to such startling information is, *"Why does the breach cause them to fail?"*

Here's why:

- It is too costly to fight the suits brought on by the end victims' that come after a business for not taking "reasonable care" to protect sensitive data. For example: in the Verizon 2015 Data Breach Investigation Report, the cost for a breach of just 100 records could range from a few thousand dollars up to $555,600.00. How many small and midsize businesses can afford that? Not many.

- Few businesses are able to recover from large amounts of cash disappearing from their bank accounts. Banks won't and don't have the same protections for business accounts as they do for consumer accounts. When your balance shows the money is gone—it is almost always gone for good, which is definitely bad.

- When businesses have data stolen, it is very challenging. . . if not impossible, to regain trust fast enough to "weather the storm." Consumers grow more educated in data theft with every story out there and are likely to go to the business that has better policies in place in the future.

It has been proven time and again through numerous Internet security studies, that all businesses are targets for cybercrime if they don't make an effort to protect themselves from it. However, when asked about it, most businesses believe they are adequately protected, which doesn't add up with the fact that 87% of them do not even have a formal written network security policy or practice in place. Remember, cybercriminals have access to these same statistics! And a cybercriminal knows how to find the "paths of least resistance" to capitalize upon.

How big is your business? How secure is your network? These questions will say a great deal about what your chances of becoming a victim to a cybercrime are. As someone who helps small and midsize businesses prevent this, the focus of my services is helping businesses reduce their chances of being breached. There is never a 100% guarantee; however, there are very effective strategies to help your business not become that next headline and/or statistic of failure. That is the point that we can both rally and educate from.

WHY WOULD A THIEF WANT *THAT?* SIX THINGS YOUR BUSINESS HAS THAT THIEVES WANT.

Belief and opinion are the biggest hurdles in implementing effective security that can help prevent an attack.

I remember growing up and hearing people say, "One man's junk is another man's treasure." For businesses, what they perceive as something of "no value," can be extremely valuable to a criminal. They will maximize it and expose it, giving their selves a pretty sweet deal while the business and its customers suffer. This likely disturbs you to your very core, but it doesn't disturb the perpetrator at all.

As a business owner, knowing what you have to offer that may be more valuable than you believe is the first step. There are 6 specific areas of data that are considered the jackpot for cybercriminals. If you know what the gold is, you'll know how to protect it better.

1. **Banking credentials**
   Think about your payroll accounts and the abundance of information that is in them. A thief will not hesitate to figure out your banking credentials and piece them together, which will give them the ability to impersonate an authorized user on the account. Then—in a matter of a minute—the payroll account is drained. <u>What would you do if your payroll account was suddenly emptied the night before payroll processing</u>?

2. **Sensitive data from customers, vendors, and staff**
   Credit card numbers, Social Security numbers, and other data that

help a thief take over someone else's identity are valuable pieces of information. In the cyber underground, they can go anywhere from $10 to $300 per record, depending on its value. <u>Does your business have any of this type of information stored on technology of any sort</u>?

3. Trade secrets

Entrepreneurs and innovators work hard, many creating products and services that become a part of all our futures. Along with these exciting challenges comes valuable information and data such as: secret formulas, design specs, and well-defined processes. There is a market out there for this information, because some people want to shortcut the path to success by copying those that paved the way. <u>Are your ideas and processes safeguarded from thieves</u>?

4. Email

It's hard to imagine that an email account could be of real value, but there is information on there that cybercriminals love. Here are some numbers that one prominent credential seller in the cyber underground can get:

- $8 for an iTunes account
- $6 for accounts from Fedex.com, Continental.com, and United.com
- $5 for Groupon.com
- $4 for hacked credentials to hosting provider Godaddy. com, as well as the wireless providers Att.com, Sprint.com, Verizonwireless.com, and Tmobile.com
- $2.50 for active Facebook and Twitter accounts

If your inbox was held for ransom, would you pay to get it back? If your Webmail account got hacked and was used as the backup account to receive password reset emails for another Webmail account, do you know what would happen? The result would be that an attacker could now seize both your accounts! And here's a startling fact: if you have corresponded with your financial institution via email, chances are decent that your account will eventually be used in an impersonation attempt to siphon funds from your bank account. <u>Have you ever conducted any personal</u>

business on your email that you don't want criminals to have access to?

5. Virtual hiding places

Using your unprotected network to launch attacks against others—perhaps one of your top clients or vendors—is a favorite technique for cyber attackers. They will expose the weakest link to their end target and literally "work their way up." They start with a smaller company that does business with a larger firm and may have access to some of its passwords and accounts due to the type of working relationship. Then the cybercriminal finds their way into that system and starts to extract the data that they desire. They may also infect the small business' site with malware. Then when larger corporate clients and vendors visit the infected site, the malware secretly attacks that person's computer and infects the organization. This is known as a watering hole attack. If you were attacked and it impacted your clients, would they understand?

6. Your reputation

The higher up the scale of success that you go to your peers, the more a few of them may desire to see you come back down a bit and "make room for someone else." There are unscrupulous competitors out there, and also disgruntled employees. Today, targeted reputation damage is a serious concern for small to midsize businesses. In fact, damaging attacks, whether it be data theft or destruction by rogue employees, has moved up to the third leading cause of loss according to NetDiligence® 2013 Cyber Liability & Data Breach Insurance Claims -- *A Study of Actual Claim Payouts.* Do you rely on your reputation to help drive your business?

Most everything that a business has access to using technology, whether it is to either retrieve or store information, is of value to someone who has made a career out of attacking businesses for their own malicious gain. I get that it's hard to accept this, because most of us do not think like a criminal, we think about our futures and conducting the best business we can. However, in order to know what you're up against, you really need to start understanding what criminals may see in your business through an honest and thoughtful perspective. It's a conversation best had with someone who understands the full scope of cyber security.

WHAT IS *NOT* YOUR BEST ASSURANCE FOR BEING PROTECTED AND CYBER SECURE?

Read the fine print.

With the continuous rise in cybercrime, more businesses have taken proactive steps and purchased Cyber Liability Errors & Omissions Insurance. According to Cyber Liability Insurance Market Trends report that was posted on October 2014 by PartnerRe, cyber liability policy purchases jumped fivefold between 2006 and 2013. *But is this enough?*

At first thought, many business owners may feel that having the insurance policy is more economical than dealing with all the technology that goes along with having truly effective cyber security for their business. They get a policy and it's not hard to get; however, they don't read all the fine details of that policy and then—kerpow!—it happens. Their business is attacked and they go to file a claim, only to hear that they are **denied**! What did they miss? In cases like this, the reason for the claim denial will almost always be a failure to implement reasonable care, which includes policies, systems, and management to protect both consumers and clients who've entrusted that insured business with their data. These policies are meant to complement data security measures, not replace them.

WHY CAN'T MY STAFF HANDLE THESE MATTERS?

Cyber risk, like all corporate risk, needs
to be managed from the top.

Businesses want to minimize their spending and maximize their output. This makes sense; however, not everything should be negotiable. **Isn't spending the proper amount of capital on IT needs logical if it keeps the business solvent**? Consultants and/or onsite IT staff are not always the best choice for the highly complex world of Internet security. Think of this:

You monitor and manage the continuing volatility in the global capital and credit markets, because you want to maintain an edge and stay informed on what is happening that may impact your business. And then there's the competition, which you watch closely so you can understand if they have products or services that may impact your business. This is smart business. But what if . . . you ignore the fact that a cyber breach could result in the loss of information that would cause you to lose a bid on a contract, lose key intellectual property, or lose millions of dollars because of an operational shutdown. All of that happened because you didn't pay attention and had no systems in place.

IT security specialists do pay attention to all things cyber security and understand how to put the proper measures in place. With the frequency and severity of cyber security incidents on the rise, it is a critical time for CEOs and boards to focus on understanding and proactively managing cyber risk with qualified technology support teams. What constitutes qualified? Having access to the trends as they are happening is a good start. And also, having the support staff to manage these things on a daily basis is beneficial. The full time IT guy is often spread thin and has a lot to do. When do they have time to train and stay on top of outside threats? Not as often as they should. This is why staff is not always ideal for handling things such as cyber security; plus, <u>an impartial perspective can be highly beneficial in finding the flaws in an existing security system or in creating a better cyber security system</u>.

I WANT MY BUSINESS TO BE CYBER SECURE. WHERE DO I START?

Learn, implement, and monitor.

Business is constantly changing and growing. Business owners and CEOs that want to apply their focus and energy to growth will invite in IT cyber security specialists that offer services such as what I do to make sure everything is optimal. Their time is better spent elsewhere! They will ask and evaluate the three big questions during this time, each designed to gauge what needs to be done and where vulnerabilities exist.

1. What data is at risk and who has access to it?

2. What vulnerabilities will attackers use to get unauthorized access?

3. What's the financial impact from a data breach?

The topic of cyber security has become a buzz that all business owners hear, but many ignore, swatting it away like it's a pesky detail. However, **there is no replacement or substitute for cyber security**. It is necessary if business owners and leaders have goals other than putting their business at risk.

Good, solid technology, training, policies, and procedures can greatly reduce businesses' internal and external threats. <u>The cost of putting this technology in place should pale in comparison to what a breach could cost</u>. These are lessons we should embrace and not "learn-the-hard-way" from. This is why I'm always committed to every effort that can be taken to stay ahead of cybercriminals' intentions and protecting my clients. When these efforts succeed, we all succeed.

To learn more about Kris and her team at **IT Squared Resource** *and how they are helping their clients win the fight against cybercrime, visit it2resource.com/cybersecurity*

CHAPTER 10

WHY IS YOUR PIZZA SO EXPENSIVE?

BY OWEN WOLLUM

Not too long ago a large, nationwide pizza chain was blackmailed by a group of hackers who had infiltrated the pizza chains databases and installed software that reported back to them credit card information of every sale. The hackers threatened to release the credit card information of hundreds of thousands of the pizza chain's customers to the public (and other hackers) unless they were paid a ransom of many tens of thousands of dollars. Did the pizza chain pay? They say no, but the hackers disappeared without releasing any information, so you might speculate that they did. If your pizza is a little more expensive today, well, guess what happened?

It's a very unfortunate thing when something like this happens and a business that should have been more responsible about protecting the information of its customers simply absorbs the cost and passes it on to its customers, but it can be absolutely devastating for an individual when someone steals their password.

IDENTITY THEFT

A man named David Crouse noticed small charges on his debit card that he hadn't made. Before long the charges became more and more blatant, adding up to as much as $3,200 a day. Crouse closed his account and opened a new one, only to find an unauthorized charge of $1,100 on the new account the day after he opened it! In less than six months, the thieves had run up nearly a million dollars of charges on his accounts!

Crouse was targeted with malware on his computer that captured keystrokes (including passwords). He estimates that he spent over $100,000 to try to salvage his finances. His credit was ruined. His identify was stolen and became public knowledge. Even after it was ruined, other thieves continued to try to use it for their own benefit.

Protecting yourself from something like this happening to you or your business involves a variety of strategies, but starts with using secure passwords.

THE ANATOMY OF A PASSWORD

A password is a combination of keys or character strokes designed to prevent unauthorized access to an account. In a sense, a password is very much like a key to a lock on the front door of your house. If you have the key, you can get in. If you don't, you can't. It is different in that it is not a physical item but a piece of knowledge so anyone who gains access to the combination of characters you have assigned as the password will have access to your account.

Are passwords safe? Yes, and no. A good password will make whatever account it is connected to much safer than a bad password, but if someone wants access to your account bad enough, they will probably be able to get in. While a good password might keep your neighbor or ex-wife out of your retirement or bank account, if the U.S. Government (or for that matter the governments of North Korea, China or Russia, or any dedicated private hacker) wants access bad enough, and are willing to apply the resources necessary to break in, they probably will.

There are alternatives to passwords, but they are often used only for extremely sensitive information. There are devices that can be attached to a remote computer that read a fingerprint, scan an iris, or require you to enter a code generated from an electronic device you keep with you. These are all more secure than simply protecting an account with a password, but they also add levels of complexity and expense so they aren't often used.

THE BEST PASSWORDS

If we have decided to protect an online account with a password, the absolute best password we could use would be a long string of random characters that includes both upper and lower case alpha characters, numbers, and special characters on the keyboard. Something like this:

GGlf[s:dlW($)[98e33hod[mDQ?v5Z7VQ6s^iY.+#0%]Z:qT[zPUAJ1FWmx9-3(

However, there are a few problems with this strategy. The longer and more difficult the password is, the more secure it is, but the more difficult it is to remember. In passwords that long, it is much easier to make a typo, and then have to start over. Provide the wrong password enough times and many online accounts will lock you out of the account, either for a period of time, or until you call in and verify who you are.

Many online accounts have a limit on how many characters you can use for a password, and which characters they will accept. For example, your bank might only allow a password of 24 characters and not allow the use of any special characters. Another online account might allow special characters, but limit the password length to 16 characters.

A strategy that some people use to create the best type of password is a password manager that automatically generates complex passwords and remembers them. Examples of popular password managers are *Dashlane, LastPass, RoboForm* and *Zoho Vault*. Some will even have the intelligence to realize what web page you are on (for example your brokerage account) and plug in the correct user name and the password for that account.

While this strategy has its benefits and might seem like an ideal way to protect your accounts, it does have its shortcomings. As noted before, different websites have different requirements for passwords. Your password manager might create a password that's too complex for the website or doesn't meet its requirements in other ways. There is also the issue that you can access your manager with a password. All your passwords become vulnerable if someone learns the one password to your password manager.

Always do your due diligence and use a reputable account manager because there have been cases where hackers have created password managers that actually provide them with the information you store in it!

PASSWORD VULNERABILITIES

Here is a list of ways people can learn your password:

1. **Physical clues:** It is far easier to determine the password to the account if they have physical access to your account than trying to break into it from a remote location. While it is hard to believe, people put sticky notes with the password written on them on their monitor, under the keyboard, on the desk, and other obvious locations. Another common error people make is writing their password on the back of a business card they carry with them in their purse or wallet.

2. **Guessing simple passwords:** If your password is too simple – a hacker can guess their way in. Many people use simple passwords like "password" or "123abc" or "opensesame" that are simply easy to guess. Never use dates such as birthdates, anniversaries, or names of pets, children, etc. While it takes a little more work, if someone knows you well enough and is willing to work a little to get into your computer or account, they will probably be able to gain access in fairly short order.

3. **Password cracking programs**: There are programs available, like Orphcrack that can be used to determine the password of a computer account. They typically require physical access to your computer and some time to work, but if your password is not complex enough they will eventually be able to provide the user with your password. These programs can be quite helpful for IT professionals trying to rescue the data off of a computer that the user can no longer remember the password of, but they are also valuable tools for those with nefarious intentions.

One of the advantages we have of protecting an online account, is the practice of locking accounts (either until someone verifies who they are or for a period of time) after a certain number of failed attempts. This puts a heavy damper on those "brute force" attempts of trying to guess what a password might be. A program like Orphcrack working on your computer with nothing to stop it might try hundreds of thousands of possible passwords per second. When someone manually tries to guess a password over the Internet or writes an automated program to do the same, that can't happen.

THE BEST OF BOTH WORLDS

So now that we know all about what <u>not</u> to do, let's explore creating passwords that are somewhat easier to remember but difficult for someone to guess. The secret is more related to the length of the password as opposed to how difficult and obscure it is.

Take for example, the following two examples:

*$tK-'~9

four#Pelican

You might think that the first password would be safer than the second. But according to the website: http://howsecureismypassword.net, the first password can be cracked by a desktop computer in 20 days, while the second would take 64,000 years by the same desktop computer! As you can see, the first is hard to type, and nearly impossible to remember.

But if you can combine the two, you have a more secure password.

Take for example the following password:

adlibbbiiwawm1905

howsecureismypassword.com reports that it would take a desktop computer 2 billion years to crack this password. [Note: obviously 64,000 years, or even a year, sounds like a very secure password. The point to remember here is that the longer it takes, the more secure the password.]

While it may seem like this password is composed of a long string of characters that would be difficult to remember, it's actually very easy for me since it represents the first letter of each word in a phrase. In this case, it's a line from "If I Were A Rich Man" from the movie and stage presentation, *Fiddler on the Roof*. The 1905 appended to the end is the year that the movie/play was set, and is a critical part of the password as taking it off drops to 19 years the time it would take a desktop computer to crack it!

Oh, and don't bother trying to log into any of my accounts using that password! I've since changed it.

CHANGE THEM OFTEN

It's critical that you change your password frequently. How frequently? That's up to you, but the more often the better. Of course it's not a fun task. According to the results of the 2012 Online Registration and Password study, 38% of the respondents would rather undertake household chores, like cleaning the toilet or doing the dishes than have to create a new username and password! So we have to force ourselves to do it.

Some websites will require that you change your password frequently. Others, don't require that you change it all. I try to do it every six months or so on my critical accounts like my banks and retirement accounts.

Another important practice is to use different passwords for every account that you have – especially the important ones. If you don't, and someone does gain access to one of your accounts, then they'll have access to ALL of your accounts. This can be especially devastating if they gain access to your email account. You probably have your bank and other vendors send you statements and notifications to your email account. If someone who intends to steal from you has access to your email account and you're using the same password for all the rest of your accounts, then you might as well just turn everything over to them. They'll simply work their way through your email and find out who is sending you email, then log into the accounts.

You may have the strongest passwords in the world, and change them regularly, but if you don't take precautions to protect your computer in other ways, it won't matter. If a virus that includes a keystroke logger takes hold of your computer, passwords won't mean a thing. Each time you log into an account, the keystroke logger will send the URL, the account name and the password back to the hacker who created it. You can bet if there's much money in your bank account they will be finding ways to grab on to it and spend it for you.

There are several ways you can protect yourself from viruses (including keystroke loggers). They include:

Always keep your operating system patches updated. You can set your computer to do this automatically whenever patches are released. Doing so is an excellent practice.

Have a good anti-virus program installed and make certain that it is updated regularly. Again, you can set it to automatically install updates whenever they are released and you should.

Always install updates to third party programs that you use (especially common programs like Adobe Flash, Adobe Reader, Java, etc.). Just make sure that when you install them you don't install the extra "bloatware" they try to install automatically (you have to uncheck the box for them) or you may eventually end up with a virus just from updating one of them! In my experience, the ask toolbar is one of the worst offenders of piggybacking potential dangerous viruses with it.

Stay away from websites that aren't absolutely necessary. Viruses are prolific on sites that include pornography, other images, and game and music downloads. If you're going to visit sites like those, make sure you know you can trust the site.

Never click on a link someone sends you through email unless you are absolutely certain it is safe. If you have any question, call or text them first to make certain they actually sent it to you and that it wasn't some virus working on their behalf.

MULTI-FACTOR AUTHENTICATION

You can always augment the security of your password by using additional forms of protection. With some vendors you can either get a device that will give you a code that you have to enter along with your username and password each time you want to log on. The code is in an electronic device that you can attach to your keychain and generates a unique code tied only to your account each time you press the button.

Another way to provide additional security when logging into online accounts is by having the vendor text you a code to your cell phone number (which you have previously registered and authenticated with them) which you are required to enter along with your username and password.

Both of these methods are different approaches to what we call "multi-factor authentication." To use it you have to both know the username/password of the account and have possession of some kind of additional authenticating device. Very few vendors provide this as an option, but if you have the opportunity to use it, you should take advantage of the opportunity on all important accounts.

After reading this chapter you may think that it's impossible to safeguard your accounts with passwords. In a way, that's true – if someone wants access to your information bad enough, and they have the resources and time available to get it, they will. But in most cases, following the advice I've shared in this chapter will keep you quite safe.

About Owen

Owen Wollum helps owners of small to medium-sized businesses and non-profit organizations leverage their technology to grow their companies and make them more profitable. He is the founder and CEO of **Dependable IT Support** which abides by the philosophy that technology is simply a tool to help an organization, and should not be the focus (unless your company is a technology company, of course). The less you have to worry about the technology that enables your processes, the more you can focus on whatever it is that makes your organization great.

Owen has never been published prior to this book (unless you count articles in his high school newspaper!) but has taught computer classes for many years, and enjoys providing easy-to-understand examples and analogies to help people quickly grasp complex topics.

Owen has worked with technology companies since the dawn of the microcomputer age in a variety of roles from consultant, strategic and financial advisor and "virtual IT director," to programmer, database guru, teacher, and security analyst. He is a graduate of Pacific Lutheran University, and enjoys hiking and sailing.

CHAPTER 11

RAISE YOUR ANTI-HACKING IQ: BEATING HACKERS AT THEIR OWN GAME

BY AHSUN SALEEM

Everyone is going to get hacked - it's just a question of how quickly we respond, or go on the offense.

I've been in the IT industry for 15 years, and I've seen many friends, family and clients get hacked. Many people think that hackers are little kids who are just having fun, but the reality is that hacking is a very lucrative business, which means that you are up against highly-motivated criminals. They are almost always a step ahead of everyone, making it hard for consumers or businesses to effectively fight back. Many of us are busy with our career and personal lives and therefore aren't vigilant in monitoring our online presence. In fact, we take for granted that our accounts and data are safe because we set a password for them. Hackers are after your personal data, and the consequences of a hack are disastrous – at the very least you'll be spending a lot of time trying to clean up the mess left behind by hackers, and at the very worst, you can spend months trying to get your identity back.

There are ways to minimize the risk of getting hacked, and I want you to know what you can do to better protect yourself and your data against hacking attacks.

WHAT DOES IT MEAN TO GET HACKED?

It's common for someone not to realize that they've been hacked.

A "hack" occurs when an unauthorized person retrieves your personal information to use for their own gain. *It may be so subtle that you don't realize that it even happened, or it may be so bold that your world is instantly disrupted—a personal catastrophic event.* Hackers are tricky, and aside from someone stealing your credit card and going on a shopping extravaganza with it, it can be really hard to detect if you've been victimized. When your personal information, such as passwords, Social Security number, Drivers License number, account numbers, and other information are taken by a hacker, they don't always use them immediately. They may sell it, or else wait a bit before using it, making it even more difficult to track them down.

Getting hacked could mean:

1. Having your bank accounts cleared out.
2. Becoming the victim of identity theft.
3. Unknowingly using a computer that is compromised, making you highly vulnerable.
4. Your business' website is being redirected to another site, or is being used to attack other websites.

HOW DO YOU GET HACKED?

I. **Through your personal devices**. Your laptops, desktops, tablets, and smartphones have your personal data on them. These devices bring a level of convenience to our lives, and we gladly use them because it allows us to work more efficiently or stay connected to our loved ones; however, it does make us vulnerable if we don't take a few proactive steps to keep that valuable data protected.

II. **Through Third Parties.** Sometimes you can get hacked through no fault of your own. Many companies also hold your personal data and that data can be compromised. For example, there was a report that revealed that 1,000,000 credit card numbers were stolen off of one person's laptop! It's important to understand that **major websites are always under attack from hackers.** This means that if you have accounts with well-known companies (think banks and the major free email providers), there is a potential for hackers to gain access to your data.

ALERT: IDENTIFYING THE SIGNS THAT SHOW YOU'VE BEEN HACKED.

Don't assume it's a glitch when odd things happen in your digital world.

It's important to know the signs of a hack, because if you are not aware or chalk it up to a "technology glitch," you will not take the appropriate actions in a timely manner, which makes the situation worse. Some signs are:

1. Your friends tell you that they are receiving weird emails from your account. This is a sure sign that you've been the victim of a phishing scam.

2. Your online passwords have been changed—and not by you!

3. You receive a call from your credit card company or a legitimate email or phone alert that says there has been suspicious activity on your account.

4. There are unauthorized charges on your phone bill. This happens when malware is installed on your phone, which could happen if you don't purchase apps for your smartphone through a trusted app store.

5. You receive a notification from a company that you've conducted business with, stating that your account may have been compromised. Use caution with this one! It may be a phishing attempt. Call the company directly first and verify the email's content.

6. Your computer exhibits slow performance and/or you are receiving pop-ups about your PC being infected and demanding payment to remove viruses.

7. Browsing the web takes you to a different page than you intended.

8. You cannot access your anti-malware program, or other system tools that are available on your PC.

These eight telltale signs are sure indicators that something is happening with your computer that is compromising you in some way. Take these warnings seriously, and most importantly—take action!

I'VE BEEN HACKED: WHAT SHOULD I DO?

If you've been hacked, make it a priority to start
regaining control of your information.

Depending on the severity of the hack, it may take some time to clean up, but there are very specific things you can do immediately to stop the damage from becoming worse.

a. Change your passwords
Change the passwords to all of your accounts. Start with your email accounts as all other password reset emails will be emailed to you at the account on file with those companies.

b. Be proactive in regaining access to accounts that you've been locked out of
When you're locked out of one of your accounts—Facebook, for example—reclaim your account by following the directions provided by the specific company. Remember, if you are locked out of your account, someone may be on it pretending to be you and having access to information (personal and possibly professional) that you don't want them to have. They can damage your reputation.

c. Contact friends and family
Let people know that you've been hacked. This is for everyone's protection. Inform them not to click on any suspicious emails that they may have received from you.

d. Act quickly if your computer has been compromised
Disconnect your computer from the Internet, if possible. This step does makes it considerably more difficult to eradicate the infection, but it may help prevent further data loss or damage. Then:

- *Run an anti-virus and anti-malware scan*
  Make sure you don't run this through a new and unexpected prompt on your computer. If you don't have these programs in place, research the best programs to use from friends or an IT professional. Download the program to a USB stick and then install it on your computer if you're disconnected from the Internet. If you are on a PC, you may have to boot into Windows Safe Mode to install the applications.

- *Run Windows update*
  After you've cleaned out the offending software, make sure you

run a Windows update to ensure you have the latest patches and security fixes installed. On a side note, don't put off performing updates of your installed programs when available. Many of these updates are designed to patch vulnerabilities in the program that hackers exploit. You can schedule the updates to run automatically to make it easier to stay up to date.

- *Make sure you have a current Windows program*
Windows XP is no longer supported by Microsoft, which means they will not release any updates for it anymore. If you are still running this OS on your computer, update it to the current Windows version. Otherwise, patches (which fix the vulnerabilities that hackers exploit) will not be available to help protect your computer and the data it holds.

e. Deal with identity theft quickly

Identity theft is disruptive and usually takes months to clean up, which means that the quicker you take action when you discover you're a victim of it, the better your chances are to overcome them.

- *Report it to the 3 major credit bureaus*
You can file a fraud alert with all the major credit bureaus. This is extremely important as it protects your credit history from being impacted by fraudulent activity.

- *Request new credit cards*
Inform your credit card companies about the attack and request new cards immediately. This will prevent further fraud and remove any responsibility for fraudulent charges on your account. Consider doing the same thing for bank accounts, too.

- *Contact the authorities*
Call your local police department to report the identity theft, and provide any documents you have. From there, they can guide you to the best ways to get an investigation started and hopefully find the criminal who stole your identity—although, this is not always easy to do.

- *Monitor your monthly statements carefully*
Since hackers and identity thieves can be relentless, you'll want to pay close attention to your monthly statements. Question charges and don't assume that they may be from a spouse or too small to be significant. A lot of random $5 dollar charges can

add up to quite a big criminal payday in the end when they are not addressed.

Being hacked changes everything instantaneously and forces you to pay closer attention to your life's personal details. And it only has to happen once for you to definitely know that you never want to go through it again! Thankfully, there are highly effective strategies to help prevent hacks.

PREVENTION: GO ON THE OFFENSIVE.

*Take advantage of common preventative measures
that exist to deter hackers.*

You can make things challenging enough for hackers by adding layers of security to your online accounts. Setting up protection is everything in the pursuit of prevention. Give yourself some peace of mind by making sure you've done what you can to deter cybercrime from affecting your life.

1. **Whenever possible, do not use an obvious username**.
   When a hacker can figure out your user name, they have already won half the battle. A growing trend is for websites to have you use your email address as your username, however, email addresses are easily discovered because they usually follow a typical format (firstname. lastname@emaildomain.com/firstnamelastname@emaildomain. com).

   Instead, create a unique email address/user name that you can remember. Another option is to use multiple email addresses, one for newsletters, coupons, etc. and another for personal communication. Since many companies still do sell email addresses, this will help reduce the risk of your personal information being compromised.

2. **Use strong passwords**.
   Easy to remember passwords, such as names, birthdates, and Social Security numbers of loved ones are easy to crack. Passwords like "password" or "12345" are very common, and also easy to crack. Make your passwords reach the "strong password" marker that many businesses' websites have to better protect your information. Don't settle for passwords that are categorized as "weak" or "medium" by websites, even if they accept these passwords.

The best passwords have a random mix of upper case and lower case letters, numbers, and characters. Try using acronyms to make them easier to remember. You should also have several passwords being used across your accounts so you're not using the same password everywhere.

3. **Use "2 factor authentication" where available**.

This type of authentication requires a user to have an extra form of identification aside from a user name and password. These forms include:

- Security questions to which a stranger wouldn't know the answer
- A special PIN code sent to your phone from an app, or via text message
- Something physical, such as a RSA key fob (found in corporate settings)
- Your fingerprint

4. **Be careful about what you click on**.

Hackers use phishing schemes to capture sensitive data. Phishing emails look legitimate—like they came from your bank, credit card company, or even the government sent them—but they are not. And please note, practically no official corporate or government organization will ever ask for sensitive information over email. A warning sign of a phishing scheme is the email address is slightly off from the name of the sender/company.

Never click on a suspicious link, either. If something comes from someone you don't know and it isn't something you are expecting—do not click on it. Even when you recognize the sender, don't click a link if you don't recognize it or if you weren't expecting them to send you a link. It could be that they've been hacked!

All of the major web browsers have pop-up blocker functionality built in. Do not turn this feature off as it will help prevent attacks on your computer. You can customize the pop-up blocker to allow pop-ups on websites you trust.

5. **Keep your systems up-to-date**.

Always keep your technology systems updated. This protects you and your devices from vulnerabilities. Purchase a reputable anti-virus program for your PC's. They will automatically scan and update your device, which keeps you better protected. Set up automatic billing so the program never expires.

6. Be cautious with social media.

It's wonderful to share with others, but when social media pages become an open book, you increase your chances of becoming a victim of a cybercrime, and possibly other types of mischief, too. Enable privacy controls and only share sensitive information with people you trust.

7. Back up your data.

Hacker attacks are getting more sophisticated. They now write viruses that encrypt your data and require you to pay a ransom for you to retrieve it. Backing up your data often will protect you from these types of attack and make sure you are not held hostage to their demands.

8. Be cautious on public Wi-Fi networks.

Public Wi-Fi networks, the ones that are accessible to everyone are the hackers playground. NEVER transmit confidential and personal information over public Wi-Fi network, as a hacker will easily steal that information. If possible, avoid public networks altogether as your device is vulnerable on them.

9. Be aware of public "cloud" storage risks.

Amazon, Dropbox, and Google are well-known public cloud storage services, and hackers are always trying to infiltrate these goldmines of information. Don't store sensitive data in these locations.

CONTROL: YOU HAVE MORE THAN YOU MAY THINK.

Take control.

Getting hacked may seem like it is out of your control, but there are many things you can do to reduce your risk. Staying alert, informed, and proactive will go a long way in protecting yourself from these attacks.

Evaluate your data and how safe it is today. Helping clients defend themselves is a rewarding aspect of my career, so if you have any questions on how you can protect yourself or your business, I encourage you to reach out to me. It's time to take and keep control of your data!

About Ahsun

Ahsun Saleem is the President and Chief Executive Officer of New Jersey-based **Simplegrid Technology, Inc.** Simplegrid is a full service IT consultancy that specializes in Managed IT Services and Technology Consulting for the legal, financial and healthcare industries across the United States.

In his role as CEO at Simplegrid, Mr. Saleem sets the firm's overall business strategy and oversees policy for all business operations. Mr. Saleem oversees the firm's marketing, operations and finance departments, along with managing the firm's consulting staff. Mr. Saleem's greatest passion is helping other businesses succeed and is very involved with his clients. Mr. Saleem assists clients ranging in size from Fortune 500 firms to small boutique-size firms to shape their IT strategy through the strategic use of enterprise grade technology. Mr. Saleem has been very successful in helping his clients increase their profit by either generating revenue or reducing cost via the various IT solutions he has recommended.

From his years of industry experience and work within the healthcare services and legal industry, Mr. Saleem has an intimate knowledge of the institutional level technology governance standards, and operational risk management that business owners expect.

During his tenure at Simplegrid, Mr. Saleem has been recognized as one of the top 250 Managed Services Executives in the world by MSPMentor and has been named one of the top 150 IT executives in the world by SMB Nation. Mr. Saleem is also a distinguished speaker for the International Legal Technology Association's annual convention and at its regional conventions.

Mr. Saleem studied at Rutgers University (Rutgers College) where he received a Bachelor of Arts Degree. Mr. Saleem also studied in the Executive Masters in Technology Management program at the Wharton School. Mr. Saleem recently earned his Masters of Business Administration degree at the Kellogg School of Management, Northwestern University.

During his free time, Ahsun is an avid golfer and softball player, and loves spending time with his wife and three beautiful children.

CHAPTER 12

RANSOMWARE

BY ALEX ROMP

Ransomware, as it is known, is any type of computer malware or virus that restricts access to a computer and demands money to restore access. The first known instance of Ransomware appeared on the horizon in 1989 as the "AIDS" Trojan. But Ransomware remained relatively rare until sometime around 2006. It is quickly becoming one of the more common threats, however. Infections doubled between 2012 and 2013 alone.

Traditional malware is designed to spread to as many computers as possible. They can steal information by searching your PC, stealing passwords to websites or tricking you into giving them sensitive information. Ransomware, however, prevents you from accessing your computer or data until you pay a ransom.

Varieties include fake attempts at extortion which is sometimes called "Scareware." This can present itself as a fake antivirus program, which finds dozens of fake "viruses" which you must pay to remove, or a fake Windows activation notice, saying you must pay to unlock your copy of Windows which it claims isn't licensed properly.

One such virus included "WinLock" which blocked access to your computer until you sent a "Premium SMS" message, which ended up costing you around $10. The authors of WinLock made an estimated $16 million on that one. Others extorted money by capturing your IP address and a picture of you from your webcam which it displayed on the screen along with a fake FBI notice saying that illegal software was found on your computer. It required you to pay them with GreenDot by MoneyPak or something similar in order to "pay your fine" to access your computer again.

These "Scareware" programs and other fake extortion schemes didn't inflict a lot of damage on the infected computers. When we would run into those, we simply removed them and the user was no worse for the wear. In 2013, however, the playing field changed significantly.

In September 2013, a new type of Ransomware emerged called CryptoLocker. This malware would encrypt the contents of your computer along with any connected network drives. The program would work silently until it had done its damage, then it showed itself. It asked for $400 which had to be paid using anonymous Bitcoin currency within 72 hours. After that, they claimed they would permanently delete the encryption key resulting in you losing your data forever.

CryptoLocker infected nearly half a million users. Estimates vary widely but it's believed that the originators made $3- to $27-million from CryptoLocker. International authorities were able to eventually shut down the servers that controlled CryptoLocker. Luckily, it was found that they never deleted the encryption keys for their victims' computers. A website was setup where users could obtain a program to decrypt their computers if they had previously fallen victim to CryptoLocker and not paid the ransom.

After CryptoLocker was shut down, other variations of it started appearing, looking to capture some of the success had by CryptoLocker. The most common one of these is called CryptoWall. It worked in much the same way as its predecessor, however it used a method to hide its servers from the Internet. The ransom you had to pay to decrypt an infected computer increased as well.

A macabre variation on this theme is "Tox." This platform permitted anyone to build their own encrypting malware and even offer custom malware as a service, a sort of "Build-a-Bear" malware style. Its teenage creator put the service and website up for sale within a week due to the high demand. Look for more of these to spring up in the future.

Naturally anything this devastating will have people asking, "How would I get infected and what do I do when it happens?"

The truth is that most viruses are invited in. The user attempts to download something online but was sent to a malware payload instead. They trusted what they saw and innocently authorized it to run. Sometimes the virus pretends to be an important message, like a received fax or voicemail, or a fake receipt to a large purchase from an online store such as Amazon. CryptoLocker was typically sent via a ZIP file with a fake PDF inside that masked the executable (.exe) file carried inside. Yet others were handed off by what is called a "drive-by attack." This is a malicious or infected website that was able to exploit a vulnerability of your system in order to directly infect you.

Awareness is key. The more layers of security you have in place, the more difficult it is for the virus to reach your computer. Suggested methods include a firewall with antivirus, a SPAM filter, virus protection on the computer, and DNS filtering to prevent the malware from checking into its command and control servers (e.g., OpenDNS). Be careful when you open any email attachments or click on links in emails. Even if you recognize the logo, such as from your bank or a huge site like Facebook, the people trying to get into your system know more about you than you might think. There is a massive amount of personal information out there about almost everyone.

You want to have security even at the point where the traffic enters your network, which for most people is their router or firewall. Many people confuse the office supply router that you can buy for $39.95 and think that's a good firewall. In fact, some of the packaging even calls it that, but they are actually simple routers. Modern firewalls tend to have protection layers like those we see in what is called a UTM (Unified Threatening Management) appliance.

These firewalls have the security layers, anti-malware, anti-virus, and they perform URL and web content filtering. There is intelligence built into the firewalls themselves. The down-side is that they're quite a bit more than $39.95 and they usually have an annual subscription associated with them in order to maintain the signatures and services that keep the security filters up to date.

Another safeguard is the spam filter. You definitely need to have something watching out for malicious emails, although those are never bulletproof. One person we helped that got infected had received the email from his attorney after that email went through two separate security layers and still got in. So nothing is ever going to stop 100% of threats.

We very recently had a client send us an email which said, 'Hey, I need to be able to open this attachment. It's from one of my clients.' Well, it may have been from one of his clients, but it broke our rule. If you are not expecting an attachment, and explicitly so, I don't care if it's from your mom. You probably shouldn't open it. The same was true here.

It was from one of his clients, but it was a malicious email. His client probably had a virus on their computer that was spreading in a similar manner and it sent him a poisoned email in this case. Luckily our security services which were running on their system prevented him from being able to connect to the linked website that was in that particular email.

A lot of people don't have such security protocols in place, so they could end up with a type of malware, including some form of Ransomware. So even though some people think email viruses were a threat in the past, this example proves they are alive and well.

I mentioned DNS filtering as a way to prevent viruses from checking in with their command and control servers. Most Ransomware has a command and control server it talks to because they're extorting you; so they have a mechanism for the virus to check in with its masters, so to speak, and for you to pay the ransom.

So this is another layer that helps prevent that, because if they can't check in to get their commands and get the encryption keys, they can't encrypt your system.

These things are very cleverly disguised. There's a lot of money going into the R&D to create programs like these. Therefore, we advise you not to open email attachments or click links to emails unless it's something you're explicitly expecting. Unless someone says, "Hey, I'm going to send this document over to you," you probably shouldn't be opening it.

To prevent the "drive-by" attacks, your systems should be patched and up-to-date. When we see those notifications to update the software on our computers, human instinct is to get rid of the obstruction by clicking whatever makes sense; in this case a "Go Away" or "Remind Me Later" button. However, if you don't install these security fixes, a malicious site can take advantage of your system to install malware. To every degree we can, we don't even put our clients in the position of having to accept those things. We generally set them up to be completely automated.

The most important step you should take is to backup your computer; while it's clean. Even if you lose a few files in the interim until your next backup, it's better than losing everything you have on your system. Those lovely family photos can become a thing of the past. You should store one back-up on-site and another off-site, such as at a friend or family's house, or in the cloud using a business-friendly online backup solution. If you're running a business from your computer, your customer's information is your legal responsibility. This means you can be held liable for credit card numbers, social security numbers and more.

I tell clients and prospects all the time, "I can recreate every stitch of your network if your building burns down or anything else. One thing I cannot recreate is your data. So you need to have a backup. It needs to be monitored actively and it needs to be tested routinely. The time to test your backup is not when you're running a restore."

If your system should become infected, begin by removing the malware. It's generally a simple removal process using an anti-malware application. Then move to the recovery stage from that backup you've created while the system was still clean. Enlist the aid of an IT professional who can often only restore what was encrypted instead of having to restore everything.

You can try a decryption service if you get stuck, but it's a long shot. Security vendors collect decryption keys and you can test one of your files to see if they have the key. They will provide a decryptor application for free if so. The last resort is to pay the ransom and hope for the best. While you are specifically supporting criminals, it might be the only option you have, particularly if you didn't make a backup, or if that backup was also encrypted.

This is generally accomplished through Bitcoin with modern malware. Buying Bitcoins is not a simple process the first time. Once your "life" has been restored, learn your lesson and protect yourself from future infections. Don't think that because you're on an Apple product that you're safe; they are still highly vulnerable but there are fewer Apple systems out there so the creators stand to profit less for the same amount of work. Even your cell phone is vulnerable to certain types of malware.

This absolutely stands to affect the future in networking as the stakes, as well as the global danger, mount. It's important to remember that network security has always been an asymmetrical battle. The network security people have to be right 100% of the time or they lose while the hackers have to only be right *once* to get in and infect a system.

The future may see the current approach to security turned upside down. The current antivirus and detection packages rely on blacklisting, thus only the known and recognized threats can be prevented. In the future, the opposite may hold true where only the whitelisted applications, proven and guaranteed, will be permitted to operate. It could be possible to develop a middle ground between the two; keeping an unknown application in a virtual "sandbox" to restrict its access, or to at least monitor it until its fully analyzed, at which point it gets whitelisted or blacklisted as required.

Remember that you may not be the hacker's ultimate target, but a gateway to it. One of our clients had someone with a fake email address in the name of the CFO of the company asking, "I need you to wire transfer this money to this customer or this vendor." The controller didn't recognize it as a threat because the guy's email address was almost identical to theirs and the hacker had emailed back and forth several times so it appeared to be completely valid.

Losing control of your system and connections can amount to a shut-down of your personal and professional life. Take the time to make yourself the least vulnerable target possible. It could make all the difference.

About Alex

Alex Romp is the President of **Artech Solutions, Inc.**, an IT consulting firm located in West Des Moines, Iowa, that helps local businesses solve IT challenges and make their infrastructure support their businesses. Like many other Managed Service Providers, Artech began as a simple break-fix company, but Alex quickly realized how the Managed Service model was the future because of the great benefits it provided to his clients.

Artech's focus is primarily on managed IT services and network consulting. It has always been a service company first and foremost, and didn't push product sales to its clients unless it was something that would benefit the client. Alex puts his customers and employees first and he teaches his employees that customer service is the primary function of their jobs. Too often IT people are seen as arrogant individuals who talk down to non-technical people. Alex reminds them that the clients they're serving are probably a lot better at what they do than the technical person would be.

Alex has been working on networks since he was 11 years old – where he taught himself how networks run while he helped his father's accounting business manage their Novell NetWare system. He helped his high school and many other businesses enter the age of the Internet over the past 25 years.

Prior to Artech, Alex was the youngest Network Engineer at his previous job from the day he started until the day he left. He spent several years in that company and quickly climbed the ranks to become the Senior Network Engineer and the Service Department manager. In that position he learned a lot about businesses and technology and eventually left to start his own company in 2002.

Alex is driven to solve any problem that comes his way and has a passion for teaching others how to use technology to advance their businesses. With a family background in accounting, he understands the language of business and the numbers that drive it. He can help make your business IT infrastructure into the asset it's meant to be, instead of just another cost center.

Away from the office, Alex enjoys spending time with his wife and children. He often spends weekends working with the animals on their family farm or working on some other project that life on a small farm requires – from updating the electric in the barn to fixing fences. Alex likes sharing his passion for the outdoors with his family as well as his hobbies – like camping, hiking and photography.

You can connect with Alex at:
alex@artechsolutions.com
www.linkedin.com/in/alexromp

CHAPTER 13

ANTI-VIRUS SOFTWARE DOESN'T WORK:
STRATEGIES TO PROTECT YOUR BUSINESS AGAINST INCREASINGLY SOPHISTICATED MALWARE ATTACKS

BY CHUCK POOLE

What, that can't be right? Antivirus Software can't fully protect me? But my computer person just installed "Gee-Wizbang Antivirus version 22.5" for all of my computers. She said it is currently the premier product on the market and will provide the best protection available anywhere. She showed me all of the reviews and everyone says it's the best. So if you are right and the odds are that I will most likely be infected at some time in the future regardless of the security software I choose – what should I do?

Many people think that they are safe because of the kind of computer they have or the brand of security software they are using. But are you really protected? Are you as safe as your IT technician has you believe you are?

Consider the story of a large law firm I recently consulted with. They had a Professional IT firm handling their needs, they had a well-known Anti-

Virus, they had a security appliance filtering their Internet, and their computers were locked down to prevent infections. Unfortunately, they had been hit by major virus outbreaks no less than three times in the past year. The most recent attack was a class of virus called Cryptolocker. Cryptolocker is a type of virus that infiltrates a user's PC and then proceeds to encrypt all of the files it can find both locally and on a company's servers. Once the files are locked away, they demand a cash ransom for the key to decode them. Many in the mainstream media have not so affectionately dubbed this type of malware as "Ransomware" due to the fact that they are kidnapping your data for money. As you can imagine, the cost of this type of infection is not limited to just clean-up activities. The computer users of this particular firm, which encompasses dozens of employees, were not able to work for two days while their data was restored. This means the Firm was financially on the hook for payroll costs and most importantly the embarrassing loss of client goodwill in addition to the hard costs of IT resources used to remediate the problem.

You may be asking, *"So if a large firm, who employs many best practices and has good Anti-Virus software in place gets hit, what can I do to defend my company?"*

Antivirus software alone will not fully protect you now or in the future. Every day, hackers are discovering new threats and exploits. Anti-Virus software will always be a reactionary product that struggles to keep up. If you truly want to maximize your protection against malware, you have to implement a holistic strategy that goes beyond software and filtering technology. You must implement the "Four Pillars of Protection":

1). Education

2). Layered Security Technology

3). Monitoring

4). Backup

1. Education. By far, the most important of the Four Pillars is Education. Keeping your staff up-to-date on the latest methods the "bad guys" use to try and infiltrate your business is a key strategy. Many professional IT Support companies produce seminars, distribute videos, and send periodic email alerts explaining in detail how these rogue programs find their way into your computers, servers, and network.

The fact is, most malware infections are accidentally initiated by computer users clicking on infected email attachments, surfing infected websites, or through external media such as DVD's or Thumb Drives that are passed around from computer to computer. Having a regular training program in place that highlights, for example, the top 10 types of emails that cause infections or compromise passwords, goes a long way toward prevention. None of your employees or colleagues wants to get infected, often the only thing they are lacking is a small amount of knowledge to empower them.

My recommendation is that you invite your IT Staff or Service provider to put on a Quarterly luncheon where they can go over the latest Cybersecurity and malware threats. If you have a large staff, these can be done by department over a series of days. You can also record the presentation and send a link out to those who could not attend. Companies who have implemented these Quarterly seminars have seen a drastic reduction in malware activity. This has resulted in huge productivity gains and untold financial savings from losses that can result from being hacked. As a case in point, a large Commercial Realtor we work with who had at least one computer infected every 8 weeks, decreased their infection rate to just one infection per year after we implemented a structured security training program.

2. **Layered Security Technology**. In the opening to this chapter, I mentioned that Anti-Virus software doesn't work – but that doesn't mean you don't need it. The biggest problem with most Anti-virus software is that it is always at least a day behind. Traditional anti-Virus software uses pattern-matching technology to search for the signature of a "known" virus. But the Anti-Virus software developer can only make patterns for viruses it has seen and validated as malware. In some cases, it may be days or even months before an individual Anti-Virus vendor discovers what the bad guys have done.

It is for this reason you need to use multiple layers of protection which include: Web filtering, Mobile Device Management, Advanced AntiSpam, Behavioral Analysis, a security-oriented firewall, and if possible, an additional Anti-Virus product for a second opinion. Combining these technologies greatly decreases the likelihood that you will become infected because each vendor has its own team of engineers working independently to block the various threats.

This is much more powerful than using a monolithic single-vendor approach. I am a huge proponent of running not one, but two Anti-virus engines.

There is a new breed of Anti-virus software made by companies such as Webroot that have an entirely different approach to fighting malware. Instead of creating a "blacklist" of known-bad software, Webroot creates a "whitelist" of software that is knows to be good and updates it constantly throughout the day. If Webroot comes across a newly-installed program that it has never seen before, it starts to track all of the changes being made and alerts your IT Company. If it does turn out to be Malware, in many cases it can be "rolled back." This is a much more proactive way of dealing with emerging threats that will always remain unknown.

The other great thing about a product with a "whitelisting" approach is that it requires almost no CPU utilization, so it can run unobtrusively with another more traditional Anti-Virus program of your choice giving you the best of both worlds. My company, PalmTech Computer Solutions, manages more than 2000 desktops via fixed fee agreements. So the technology we choose to protect our clients directly impacts our bottom line. If we don't choose the right combinations of security hardware and software, we end up doing an inordinate amount of cleanup work which we have to absorb as the cost of our service delivery.

By using a layered security approach, we have saved our own Helpdesk more than 83 percent of the last two years. This not only positively impacts our clients by giving them unprecedented levels of uptime, but it also allows us to spend more time optimizing their networks for future productivity gains.

3. **Monitoring**. Having all of the best technology layered into a strong shield of protection is not effective unless you are properly alerted when there is a chink in your armor. As we have mentioned previously. No anti-malware technology is 100 percent effective. Having all of your security systems and software professionally monitored allows for better management, less downtime, and a more proactive experience. If you have proper monitoring in place, one of two things will happen:

(a). You and your IT Company will get an alert that one of your Anti-Virus programs has not checked in or has not been updated. This is important because in many cases the Virus software will disable or remove the Anti-Virus protection. Having a network or cloud-based Anti-Virus program that checks in to a monitoring server allows your IT Company to follow up and solve the problem before the infection spreads.

(b). You will receive a real-time alert that you may have been compromised or that a threat was mitigated. Even if the threat was resolved automatically, proper monitoring allows your IT Company to follow up with the user and provide extra training so they can avoid a similar issue in the future.

Monitoring is a critical security component that is often overlooked. It is imperative that you select Anti-Malware products that will send you email alerts or provide for a centralized dashboard that is watched by your staff or IT Service Provider multiple times per day. Unless you are proactively analyzing the alerts that are coming from your Anti-Malware and other protection systems, you run the risk of missing a critical message that could have saved you thousands of dollars in cleanup were the threat caught sooner.

We recently took on a client who had good Anti-Virus and other protections but they continued to become infected, and ultimately this caused several other computers to become infected. One of the biggest problems they faced was that when one of their employees became infected, they would typically get a message that "popped up" on the screen and the employee would just "ignore" the message in hopes that it would just go away. There was no one set up to receive alerts, and no culture of accountability. Once we upgraded their security software, not only were we now able to quickly react to infections before they became systemic, we were also able to immediately educate those employees who were victimized by a crafty email or rogue website. Monitoring make all the difference.

4. Backup. Even with the best technology, experts, and business practices, the day may come when you get infected. This is underscored by the rash of high-profile breaches involving Target, Home Depot, Goldman Sachs, and Sony. No one is immune regardless of your IT budget, and the number of security experts you may have on staff. It is for this reason that we never talk about Malware protection without some kind of safety net. In my opening story, I mentioned a large Law Firm was down for two days while they restored their files due to infection. The ability to quickly remove an infection and recover all files quickly has become the new standard in the professional IT world.

In the past, it seemed prudent and normal to backup your data once per day. But in an era with threats coming from all angles, businesses are deploying the latest backups systems that allow for 15 minute incremental updates and can maintain data for months or even years at a given point in time. Better yet, the speed at which these files can be restored has now come down to minutes rather than hours and this makes huge impact on the organization's productivity and client goodwill.

It is important that you evaluate your current backups systems. Any software that was purchased even three years ago may not have the features needed to combat current malware threats. And even if your backup software is one of the more modern image-based solutions, it may not be set up with enough storage to provide recovery points for weeks, months, or even years ago. Companies often refer back to historical backups long after an infection that required a restoration. It is important for that data to be available.

In the case of a Property Manager who had an infection two years ago, they thought they were safe when all of their current working documents were restored. However, when an issue arose about a dispute with the terms of a lease, they accessed a seldom used document scan and discovered that the document had been encrypted from the previous infection event and for some reason had not been restored along with the original set of documents. Their current backup could only go back 1 year because they had not opted for extra storage to create a few years of annual backups. Depending on the type of business you are in and how long you have to refer back to your source documents and scans, you should opt for enough storage and backup history to properly protect your company.

In the examples above and the hundreds of businesses I work with on an annual basis, the most important tool for malware protection is the education of your staff or team. Teach them to be vigilant about phishing emails; empower them to report ANY and ALL suspicious activities to your technical support team. And most importantly, monitor your bank accounts. Whether you're a mom managing the family finances, or a CFO overseeing hundreds of millions in revenues, keep a vigilant eye on your finances. Work with your banks to setup "suspicious transaction" alerts; get daily balances; real-time "large transaction" notifications, etc. Why? Because in the end, the goal of most malware is to steal funds from you and/or commit identify theft. And more vigilant you are about your finances, the less opportunities you give the bad guys to ruin your day…or your life.

About Chuck

Chuck Poole's entire life has been devoted to the passionate pursuit of Computer Technology. He has been enamored with computers long before you could even buy them in stores. By the time he finished high school, he was one of a handful of self-taught experts writing a specialized form of telecommunication software that could intercept and route phone calls internationally. Chuck has previously worked with many large organizations including the FBI, the CIA, the United Nations, and has deployed software in more than 38 countries for various National telecommunication authorities. Some of these software systems are still being used more than 20 years later.

Throughout Chuck's career, he has watched the computer industry grow, thrive and become more cohesive with the needs of business. He finds it inspiring that pioneering technologies are being embraced by the marketplace, however he sees the value in simplicity and believes the market is coming to realize that as well.

Chuck began **PalmTech** Computer Solutions to tackle many of the inconsistencies he witnessed in the computer consulting community. Whether it was a salesperson trying to convince a client that they needed the latest and greatest superfluous computer gadgets, or an IT consultant who was unable to complete a job or resolve an issue effectively, it was glaringly obvious that most IT service companies simply were not meeting the expectations of their clients. With PalmTech, it was Chuck's desire to change that standard and move beyond those expectations.

Through PalmTech, Chuck provides elegant solutions that are easy to learn, solve core-business problems and are cost effective to maintain. After all, business owners and managers want technology that "just works" and allows them the opportunity to be great at what they do. If your technology is not helping you increase your profits, Chuck believes it is not being implemented correctly. In guiding his team, Chuck has instilled a core system of beliefs: to organize, simplify, secure, and ultimately boost productivity. The PalmTech team begins each day with this mindset.

In this book, Chuck uses his vast expertise to show you how to keep your company safe from the ever changing threats to your company's computer infrastructure. He hopes you find this book useful in your pursuit of a more secure solution for your firm or business.

You can connect with Chuck at:
Chuck@PalmTech.Net
www.FaceBook.com/ChuckPoole
https://www.LinkedIn.com/in/ChuckPoole

CHAPTER 14

IDENTITY THEFT TODAY

BY MICHAEL DINKINS

Prior to mobile generations, identity was a simple concept. Today it constantly grows in complexity. Identity is more than your driver's license, birth certificate, and social security card. It's every part of you, including your posts on social media, bank accounts, credit cards, and more. Almost every aspect of your identity can be stolen and used against you.

The threat of identity theft is growing, along with the growth of the Internet and advances in technology worldwide. Cyber criminals steal credit and health information for monetary gain. Data breaches happen every day to credit institutions and online businesses. Criminals steal complete identities and loot finances.

Identity is not only stolen online but also from theft of physical devices (laptops, tablets, and mobile phones) out of cars, businesses, homes, or schools. From there, it does not take a lot of skill or time to produce high quality counterfeit documents that enable criminals to steal the identity of an unsuspecting victim.

The bottom line is that identity theft can happen to anyone. Technology has put an end to privacy. Computers store a lot of personal information about people, their activities, and private information. Once information is on the Internet it is no longer private. While some identity theft goes unreported, and no one knows for sure how frequently it occurs, there is no question identity theft is a huge, growing problem, which costs billions of dollars and affects everyone from individuals to large corporations.

IDENTITY THEFT TARGET INFORMATION

Cybercriminals want to steal your identity by targeting your personal information such as your name, birthdate, birth certificate, social security number, mother's maiden name, birth place, and public records for monetary gain. They then use this information and sometimes your credit information to operate and fund their illegal activities. A criminal will steal credit card numbers or bank account information for fraudulent transactions or submit false tax returns to obtain tax refunds. Or they may use your personal health information to obtain products and medications to be sold on the black market and make bogus insurance claims. A large amount of damage can be done if the wrong person gets their hands on your personal information or documents.

For example, a birth certificate is a very valuable item because it includes your place of birth and mother's maiden name. With a stolen birth certificate, a criminal can easily open lines of credit because financial institutions use place of birth and mother's maiden name as verification tools. A birth certificate can also be used to obtain a passport, a driver's license, and social security card. Someone can create a new identity using your information, even if all they have is your birth certificate.

IDENTITY THEFT METHODS

The methods for stealing someone's identity are almost endless and some are much simpler than you would think. All it can take is a stolen wallet or purse with identifiable information, such as bank records, credit card statements, social security card, or birth certificate, to steal an identity. Other times, criminals put more effort into stealing an identity.

For example, new parents sometimes wait too long to get a social security number for their newborn child and criminals seize the opportunity to apply for a social security card under the child's name. Once a social security number is obtained or stolen, a criminal can use it to apply for credit and loans, a driver's license, and even receive financial benefits under that stolen social security number.

Credit card fraud is a well-known method of identity theft and occurs in many ways. For example, information such as account or card number, expiration date, and full name can easily be found on a credit card receipt. This information can then be used to make phone and online purchases, known as "card-not-present-fraud."

ATM fraud can occur when a ghost overlay, or fake numeric pad, is used to record PINs on the ATM keypad. PINs can also be obtained remotely through hidden cameras, or in less advanced methods, by cards stolen with PINs written on them or through "shoulder surfing" where someone simply looks over your shoulder to see your PIN.

Account hijacking is a more extreme form of identity theft. If enough information can be put together about a victim a criminal can take over the victim's complete identity, known as "wholesale assumption of identity." The criminal can then do things from order replacement cards and change addresses to completely taking over the victim's accounts. This can easily be done online, especially if the criminal also has your email address.

Compliance Data Breach is occurring all too often, and serves as a method for criminals to target organizations in order to simultaneously victimize a large number of individuals. It occurs when an organization gets hacked and individuals' information is stolen due to improper, or lack of, security, such as unsecured networks, improperly monitored anti-virus software and spyware, lack of updates and patches on network, or outdated and unsupported business software.

Breaches also occur due to employee sabotage, employee negligence, or employee error. For example, I visited a dental office that did not use individual logins for their seven employees, did not have an 'Acceptable Use' policy in place, and stored all patient data on one computer in the unsecured and unlocked reception area. This is an all too common example of a business being naïve to IT regulatory compliance (HIPAA in this case) and unknowingly putting their business and patients at risk for identity theft.

IDENTITY THEFT PURPOSE

Why do criminals steal identities? The top motivation is monetary gain. Once information is stolen, criminals use the information to commit crimes and fraud. For example, criminals will use someone else's information to purchase prepaid phones or qualify for loans to finance purchases, open credit cards and then never pay for these things.

The purpose of stealing identity has evolved into more than just a single criminal or hacker stealing one person's identity. Stealing identities often occurs to enable serious, organized criminals to conceal themselves, their activities, and their assets in order to facilitate specific criminal acts such as smuggling people and using false identities to commit crime.

Identity theft has even become a method used by terrorists to fund operations. Major terrorist groups, like Al-Qaeda, were discovered operating a scheme in which they stole credit card information to make undetectable purchases of cell phones for secretive communications. Terrorists have also been known to steal identity information to create false passports and travel documents to move about without being noticed, to open fraudulent bank accounts, and make undetectable money transactions.

IDENTITY THEFT TARGETS

Anyone can become a victim of identity theft, but some people and organizations are easier targets. The following factors make business and individuals more susceptible to identity theft:

- Storing information on the web or in the cloud
- Utilizing an unsecured network
- Not using a password or using weak passwords
- Failing to use good quality antivirus products, even on mobile devices
- Using public Wi-Fi, especially for banking activities
- Not password-protecting home Wi-Fi, or using the default settings

IDENTITY THEFT RISK FACTORS

Can you pass the wallet test? If you lost your wallet, or someone stole it, do you know what personal information they would have about you? Many people have things in their wallet that can enable criminals to easily steal their identity, such as something with their signature or social security number on it, their home address, bank account numbers, phone numbers, business cards, health insurance cards, and even PINs or Passwords.

While anyone can be a victim of identity theft, certain things increase your risk. Here is a list of risk factors:

- Not having a locked mailbox
- Leaving bills and checks in an unlocked mailbox
- Receiving pre-approved credit offers and loans
- Throwing personal documents in the trash (as opposed to shredding them)
- Having three or more credit card accounts
- Using checks with a driver's license or social security number on them
- Doing business with online shopping websites, large chain stores, online auctions, catalogue orders, frequent buyer clubs, contest/special offers with prizes and electronic mailing lists

Organizations often put their employees at risk for identity theft, unbeknownst to either party. Storing employees' personal information online, in a web-based or in-house database with an unsecured or improperly protected network or server, puts employees at risk. Using social security numbers as an employee number or putting it on identification badge is another. Not doing background checks on employees who deal with sensitive information, or failing to limit access to sensitive information, is another risk factor.

IDENTITY THEFT PREVENTION

There are many things individuals can do to prevent identity theft, but businesses also need to take action. Businesses are more often becoming the source of stolen information which leads to identity theft. The following are things businesses can do to protect themselves, their employees, and customers:

- Properly train employees on how to handle sensitive information.
- Hire a shredding company and have a company policy that states all personal and health information must be shredded when discarded.
- Do not allow employees to write down credit card numbers or social security numbers.

- Create processes to prevent fraud such as checking driver's licenses when taking credit cards, writing down driver's license numbers and limiting over-the-phone credit card transactions.

- Use reputable vendors that follow industry compliancy, such as HIPAA, to take care of your outsourced needs and require them to sign Business Associate Agreements or Confidentiality Agreements.

- Hire a technology firm with industry-specific knowledge to keep network and sensitive data secure.

- Use business/enterprise grade hardware and products to run your network.

- Use at least business grade, if not enterprise level, anti-virus and spyware products.

- Ensure software updates are done continuously and all software products are the latest versions.

- Use sound-deadening devices to prevent others from overhearing and obtaining sensitive, personal health information in exam rooms or waiting rooms.

- Conduct background checks on employees.

- Use an Acceptable Use Policy and firewall that prevents employees from visiting sites that could be corrupted with viruses and malware.

- Lock down software and don't allow automatic updates; hire a company to manage updates and test them before pushing them out to your company.

- Know the life span of equipment and life cycle of support on software and hardware.

- Have a password policy that requires passwords to be a certain strength and set to expire every 90, 60, or 30 days.

Breaches can occur due to employee sabotage, employee negligence, or employee error. For example, I visited a dental office that did not use individual logins for their seven employees, did not have an 'Acceptable Use' policy in place, and stored all patient data on one computer in the unsecured and unlocked reception area. This is an all-too-common example of a business being naïve to IT regulatory compliance (HIPAA in this case) and unknowingly putting their business and patients at risk for identity theft.

To protect yourself as an individual customer:

- Ask questions about how personal information is maintained and how confidential information is discarded.
- Find out if businesses use a shredding company.
- Learn how your payroll information is stored.
- Use credit cards instead of debit cards for online purchases and require retailer to ask for your driver's license by writing "Check I.D." on the signature line of the card.
- Evaluate your professional service providers' offices to determine if they leave files out in the open, lock their computer when leaving the desk, give out patient information within ear range, have security systems, and if vendors are identifiable and have to sign in and out.
- Ask if the business performs background checks on their employees.
- Find out if the business outsources services; for example, do the cleaning people have access to customer files?
- Ask if systems are up-to-date using the latest versions of software, which are supported and continuously updated.

IDENTITY THEFT DISCOVERY

It takes an average of twelve months for people to even become aware of identity theft. Warning signs include:

- Missing documents such as driver's license, passport, credit cards, bank statements, utility bill, or checks.
- Problems with mail such as statements, bills, or USPS shipments not arriving or no mail of any kind arriving.
- Stolen purse or wallet or home burglary.
- Credit card charges or withdrawals not made by you.
- Receiving bills or statements that are not yours.
- Calls or letters approving or declining credit, for which you did not apply.

- Receiving credit cards that you did not open.
- Being contacted by debt collections agency for debts you did not incur.
- Applying for a loan, mortgage, or credit and getting turned down for reasons that do not reflect your true financial position.
- Credit report showing information about accounts that you did not open.
- Criminal or civil action is taken against you for things you know nothing about.

IDENTITY THEFT RECOVERY

Once the identity theft has been discovered, then what? First of all, do not panic. Prepare to take action to collect, report, and record. Collect your information and report the suspected identity theft immediately. Most likely, you will not be responsible for the fraudulent activity, but you may be if you don't report stolen information right away.

Be sure to do all of the following:

- Freeze or cancel accounts and open new ones
- Request a transaction history of all your accounts
- Change password on all accounts
- Replace debit and credit cards
- If necessary, reissue social security number
- Work with the police and financial institutions to clear your name
- Contact all three credit bureaus to dispute any fraudulent information and continue to periodically check your credit by requesting a report

As part of your reporting, fill out the Federal Trade Commission's Identity Theft Affidavit. This can be found on the Federal Trade Commission website (https://www.ftc.gov/). Write statements to your financial institutions stating that you did not authorize someone to use your account or credit card to seek money, goods, loans, services, or credit. Also state that you did not receive any benefit as a result of fraudulent activity on your accounts or credit.

After you report identity theft and collect your information, it is time to begin documenting. Document what type of information was stolen and when. Document if you reported missing information or fraudulent activities to the police. Keep records of all phone conversations and communication with every agency and financial organization. Write down the date, time, and name of person you spoke to. Include what you told them, what they told you, what actions you took, when you did it and where you sent information, including phone numbers, addresses, and emails. Get anything you can in writing.

Save all notes, letters, and emails regarding the identity theft organized in files by date or company. Make copies of bills, bank statements, and police reports. Keep originals of everything and send only copies of required information. Keep originals until the case is over, or even longer, as it may be needed for future reporting. During this process, log all costs and time spent. If the thieves are caught and convicted they may owe you damages.

If you do not have a prepared Emergency Fraud Kit, which includes a Fraud Affidavit, copies of your Credit/Debit cards, birth certificate and social security card, contact information for your financial institutions with the account numbers, and insurance policies. Prepare one now. It's easy. Put all of the above information into a notebook and store in a safe, locked place.

IDENTITY THEFT SUMMARY

Identity theft continues to grow in frequency and severity. While this is disturbing, it doesn't have to happen to you. If you do everything you can to protect yourself and your business, you have a good chance of avoiding identity theft. Protecting yourself and your business includes taking all the common sense measures discussed in this chapter and having proper cyber security policies and practices in place.

About Michael

Michael E. Dinkins is the Founder, Owner and President/CEO of **Solutions Unlimited** (founded in 1994). At Solutions Unlimited, a Business-to-Business solutions provider, Michael delivers cutting-edge technology, Security, Cabling Infrastructure, Telecommunications, and Software Application through his 21 years of field experience and skilled support team.

Michael, the youngest son of Raymond and Marjorie Dinkins, was raised in his hometown of New Castle, Indiana along with his four siblings. After graduating from New Castle Chrysler High School in 1984, Michael earned his Associate Degree from Indiana Vocational Technical College of Muncie in Applied Sciences – majoring in Electronics, Technology and Digital Computers.

As a BICSI-certified Registered Communications Distribution Designer, Michael is qualified and proficient in designing, installing, and certifying integrated information transport systems and related cabling infrastructure. He keeps his certifications current and stays up-to-date on industry trends by attending numerous training seminars and Information Technology and Communications conferences.

Michael's passion for innovation and technology led him to open his own technology company dedicated to helping people and businesses prosper by providing technology solutions that work for their business. In addition to Solutions Unlimited, Inc., Michael operates a sister company in his hometown that provides residential services. Michael strides to provide the industries best products and services to keep up with the rapidly changing technology.

As an active member in his church, Michael strongly believes in helping others and supports his hometown community. Michael continually makes investments in his community through technology projects. All those who know Michael would call him an IT Legend and is better known as the "IT Genius." Michael currently resides in Lakeland, Florida with his wife Donna Kay Dinkins who serves as the Vice President of Solutions Unlimited.

CHAPTER 15

BUSINESS CONTINUITY

BY ROB T. RAE

There is a quiet little fire station in the suburbs of Dallas, Texas. About 5,500 square feet, with a couple of bays, some newer trucks including the pride of the station, a 2014 Peirce Velocity Fire Engine. Although fire stations are not like most business, meaning they don't offer traditional retail or manufacturing products or services, they too have IT needs similar to that of any small business and those needs are far from their core business and passion.

What excites fire fighters? It's not IT equipment or noisy servers. Five or six-year-old technology is just fine, and is usually jammed into the back of a closet because of the amount of that noise it makes.

One morning, while polishing up the bright, new red truck one of the fire fighters noticed a burning smell. Tracking the smell down quickly, he discovered that there was smoke coming from the "IT closet", specifically, their tired old server. IT obviously not being their forte, they contacted their solution provider who told them to immediately shut it down and he'd be right over.

When he gets there, he discovers that the smoke was coming from their old, tired server – something that he has been warning the client about for years. The ongoing response he got was that they didn't have the budget for it and they'll take care of it next year. For the fire fighters, buying new helmets or gear was greater priority, rather than spending it on a new server. While having state-of-the-art fire equipment was important, they really did not understand the value of having state-of-the-art IT equipment.

Like any business, even a small station will have at least a server and four or five workstations. Their data is critical and includes things like staff schedules, incident/vehicle/equipment reports, photographs, etc. Software tools allow them to complete these reports, most of which are actually mandated by the federal government and must be recorded and reported properly, each month. Like any other business, their data is critical to them doing their jobs efficiently and properly. However, like many small businesses, they did not understand how important keeping it updated was – until everything malfunctioned.

Just a few short years ago; this incident would have gone very differently than today. With cloud technology and other more affordable technology innovations over just the last three years, this pretty serious issue is now able to be somewhat minimized, specifically keeping the station up and running and fully functional while their IT issue is resolved. Unfortunately, they were relying on information stored on onsite, outdated hardware.

Let's pretend for a moment that it's 2010, and this exact scenario happens. The Fire Station calls the IT provider and the IT provider tells them to shut down the server. The provider needs to go to the fire station, which takes_time. The provider troubleshoots the issue, which takes more time. They identify that a new server is needed. Fire stations, like most businesses, don't have cash lying around for unbudgeted items. Out if site, out of mind. They figure lightbulbs and toilet paper into their budget, but it never occurs to them that IT updates have to be figured into the budget yearly, if not monthly. Similar to unexpected expenses happening to your car or home, you need to find the money. Finding the funds or raising them can take time. Once the funds are secured, the server is ordered and is shipped to the fire house, which again takes time. In this scenario, the fire station needed to wait for the next budget meeting to find the money. Even with approval and a rush order the shipping and installation could take weeks. What are the staff supposed to do in the meantime in order to get their daily and weekly reports done and submitted?

Two full weeks later, the new server finally arrives. The fire station is down with no access to the Internet during this period. They had no access to their books, or to their critical applications. The amount of time they lost is easy to measure; however, the amount of money it costs them is more difficult to measure. How long can a small business stay down before it starts affecting their bottom line?

Business continuity is not a new concept. What is new is its availability to the SMB business owner. Only a few short years ago, the ability to keep your business operational during downtime, usually caused by disasters, was a costly insurance policy that only larger, Fortune 500 companies could afford. What was available to the SMB market at that time? Nothing but risk. Risk that a disaster could put you out of business if you lost all your data. Well, not anymore.

Surprisingly, the ability to keep your business up and running around the clock has been around for a dozen years. This technology is available and can reduce or even eliminate the cost of downtime, however there's a major barrier to buying. Historically, the cost of these solutions are out of reach for the SMB market. The question SMB owners have to ask themselves is: How much is it going to cost their company if the system goes down?

The average SMB will lose 17.82 hours of downtime per year due to disasters. Disasters can come in many ways. There are natural disasters, acts of God, fires, floods, tornados, hurricanes, etc. These are the most common disasters that most people think of when discussing data protection. Interestingly enough, 97% of all disasters are not natural at all. The majority of all disaster scenarios are the result of human error. Human beings are the worst things that happen to technology.

We have all seen that mysterious email and wondered what is in the attachment – naive to what is actually happening in the background. Over the last couple of years, Ransomware has become more and more popular. Ransomware like Crypto-locker and Crypto-wall are changing the way in which hackers invade your life. No longer is it about causing havoc or stealing data, cybercrime is at a whole new level with Ransomware and the collection of millions and millions of dollars in Bitcoins.

Data theft, loss and Ransomware are amongst the most dangerous disasters out there, forget about the acts of God. Not only protecting your data, but being able to keep your business up and going has never been more important. That is why virtualization is critical. Being able to fail-over to a secondary device that will take the server's place while the issue is resolved is the critical change to today's technology. Being able to take time to solve the issue, human or act-of-God-related also allows the peace of mind that the issue can not only be solved but that you can remain profitable while it is.

Remember the days when IT was pretty simple?

LIFE USED TO BE EASY...

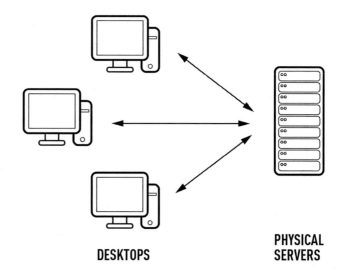

DESKTOPS

PHYSICAL SERVERS

As we continue to see more complexity in our IT environments, the threats to our data also become more complex. Security and protection have never been more important...

...BUT THEN IT GOT COMPLICATED

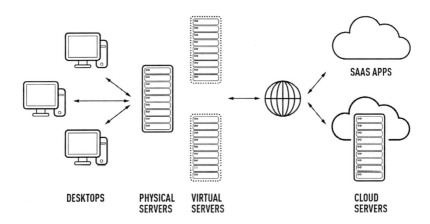

| DESKTOPS | PHYSICAL SERVERS | VIRTUAL SERVERS | | CLOUD SERVERS |

Now let's think about the act of God for a moment. Although rare, it does happen. Recently, a small restaurant had a water heater burst. Not a huge issue, and it was nowhere near the kitchen that they would dearly need to keep the business running. However, the server that they kept along with a couple of workstations were next to that same water heater. Water and technology don't mix so you can imagine the damage to their small network.

Similar to the fire station, they have schedules and reports and financial information, including payroll, that they keep on that server that were indeed critical to keeping their business going. With the server gone along with the fail-over device (their virtualization backup device), what were they to do? This is where the cloud came in. Cloud virtualization is being able to not only store copies of your business in the cloud, but the ability to work, change and operate on that image even though the original devices are no longer functional. The restaurant now has the ability to keep working, reducing their RTO and RPO or cost of downtime.

Let's look at a more serious scenario. In 2012, Hurricane Sandy skipped right over its expected path of hitting land in Florida, Georgia, South Carolina and even North Carolina. It skipped right over all the usual paths that hurricanes take and instead came ashore in New Jersey, New York and New England. The second most financially costly storm in

American history only after Katrina. The key is that in the northeastern part of the US, storms of this magnitude are just not expected. There were thousands of homes and business with serious water damage. If you had not heard – water and technology don't mix.

There were thousands of servers and workstations that were destroyed and became a pile of silicone garbage. It would take at least three months for some of the businesses to get back on their feet. But regardless of the disaster, how many businesses can have their doors closed for more than three months? How many of their employees will still be there to work after three months of unemployment? How many of their customers would go and find products and services somewhere else rather than wait three months? How many businesses would still be in business? FEMA reports that 80% of businesses in a devastated area that do not have a proper continuity plan will go out of business.

I was just in Hays, KS to visit one of my customers. They tell stories about how a tornado hit less than two hours south of them back in 2007. Tornados are common in this part of Kansas, but this was an EF5 tornado that just hung over the top of Greensburg for hours – literally grinding the whole town to dust. The last thing on everyone's mind that day was data protection. However, when the storm cleared and the dust settled people would need to get back to normalcy. Homes and businesses would need to rebuild. Business continuity planning is something that most businesses in Kansas have thought about and implemented. It's a part of starting or running any business in that area of the world as a result of the disasters that do happen.

I've mentioned recovery time objective (RTO) and recovery point objective (RPO) and have created a simple calculator for SMB's to be able to calculate how much downtime costs (you can get it by emailing: downtime@datto.com). Take a simple shop with a small data set and a small number of employees that are impacted by downtime. When you average out how much downtime an SMB will have over the next 12 months, even the smallest data sets and time down will cost into the tens of thousands of dollars. Continuity solutions designed for SMB businesses which include hybrid backup (a local copy and a copy in the cloud) as well as hybrid recovery and hybrid virtualization (if the local device is gone as discussed previously) is a fraction of that cost.

Don't live in hurricane ally – don't think you are safe. Remember that humans account for 97% of all the disasters that networks will suffer. A proper continuity plan and solution is both affordable and necessary if you want to keep your business up and going. With all the challenges of running a viable business and keeping it going, don't let data loss be the reason you end up out of business. You can remove that as an issue altogether.

The key to successfully riding out a disaster, in the many forms they come in, is proper planning. That means not only having a disaster plan in place but a proper recovery one too. Investing in the right technology that not just backs you up, but eliminates or reduces the downtime along with expediting the recovery, is well within your reach and budget. Have a replacement plan and keep your technology running smoothly. Would you buy a car and NEVER change the oil?

Top five things that every company should consider in their plan:

1. Define Objectives and Expectations. How long can your business afford to be down? How much data is your business willing to lose? Setting clear expectations within your organization and service level agreements with your solution provider.

2. Who is your disaster response team? The team should consist of members with specific roles. For example, employee safety, facilities management, and customer communications. Who are you going to call?

3. A communication plan. In the case of a disaster you'll need to communicate with many stakeholders. Your employees (onsite and offsite), your customers, emergency services, your customers and vendors and maybe even PR/media.

4. Technology infrastructure. Is your business set up properly to ensure that in the case of a disaster your data, files, and information are secure and accessible?

5. Post-Disaster Agility. What are the steps you need to take internally, and with your solution provider, to get back to normal business function?

We've all heard the expression "don't fix what isn't broken." *This may be true in most cases; however, when it comes to technology, "don't fix what isn't broken, but be prepared for when it does break," is wiser advice.*

About Rob

With over 20 years of experience in IT, Rob T. Rae has become an expert on data protection and specifically using technology for business continuity. Although Rob started his career in finance, his passion for technology drove him to join Compaq Computers back in 2000. Compaq was eventually acquired by HP where Rob spent a few more years before looking for "what's next." This is where Rob discovered the Internet of everything and the evolution of the traditional solutions and solution providers to newer technology like the cloud.

Rob joined Level Platforms (later acquired by AVG Technologies) in 2006, a network remote monitoring and management technology designed for the SMB market to manage their network infrastructure with a single pane of glass. In 2013, Rob joined Datto, one of the fastest growing technology companies in the world. As *Vice President of Business Development for Datto*, one of Rob's duties includes managing a team of evangelists, teaching the world about data protection, business continuity and the true cost of downtime.

Rob has been recognized by several IT publications, including *Channel Partners* Circle of Excellence Award, a five-time winner of MSP Mentor "Top 250 people shaping IT" award and two-time winner of CRN's Channel Chief Designation.

You can connect with Rob:
rob@datto.com
www.linkedin.com/in/robtrae
www.twitter.com/RobTRae
www.facebook.com/rob.t.rae

CHAPTER 16

CLOUD COMPUTING AND CYBERCRIME: KEEPING DATA SAFE IN A VIRTUAL WORLD

BY RAJ KOSURI

Cyber security isn't just a product; it's a process, too. The process needs to be monitored and viewed on a daily basis.

When it comes to security solutions against hackers, we don't have a supply and demand problem. **Our problem is one that involves trying to keep up with the fast-paced and ever- evolving world of "hacker know-how."** These criminals never stop working and trying to beat the system. That's why organizations that leave their data protection to chance are likely to come out of that risky gamble on the losing end.

Attackers are constantly unleashing new viruses and becoming more sophisticated on how to infiltrate networks. They rely on enterprises not understanding the latest threats or the trends in cyber security. **What's the best way you can force hackers to give up on your data that's stored in the cloud**? Start paying attention! Businesses that make efforts to stay ahead of cybercriminals have an advantage. There are steps you can take—processes you need to follow.

Over the years, I have made it a priority to think like an outsider. Why? So I can evaluate just how these hackers think and act. And then—I act, delivering organizations the information they need to enhance their data center defenses and fix their vulnerabilities so they can prevent network attacks. **It's a new approach to deterring crime in the virtual world that relies on the 3 P's: People, Processes, and Product**. *It's a combination of these three things that gives businesses an edge.*

WELCOME TO THE NEW AGE

People are the weakest link in most organizations before we come in. After we leave, they are informed and aware—able to deter cybercrime instead of unknowingly allowing it in.

We have made some progress against hackers. The traditional methods they used to use are seldom effective now. This has led to a shift in their targets. **Today, instead of focusing on companies as whole, hackers focus on one or two administrators that could give them access to an entire network**.

To dissuade attackers requires leadership from security professionals and leaders. They must proactively monitor and evaluate:

a) Major IT trends

b) The threat landscape

c) The security market

Through these three identifiers, strategic business initiatives can be implemented and security experts will become trusted advisors. It's a collaboration that reiterates the value of a proactive approach to securing a company's data.

In our current business culture, it's common for security leaders to be short on resources that help them do their jobs effectively. There is always something—an existing fire to put out or updates to be done. As a result, developing forward-looking insights about an organization's future are put on the back burner—until the organization gets burnt by an attack.

The business that understands how thinly spread their in-house IT department is or that their contracted IT company is not as advanced in cyber security as they should be, will seek out someone like me— an expert in cloud computing and cybercrime. Where others may "talk the talk," I am able to show a strong security perspective and have the knowledge that has earned me a seat at the C-suite table.

FIVE CYBER TRUTHS YOU SHOULD KNOW

Regardless of the business, there are certain truths
that apply to them all.

All businesses—regardless of what services they provide or their size—can take away certain things about cyber security and its importance to their operation. Knowing these things allows for strategic, informed evaluation and implementation of essential processes that are necessary for the enterprise.

1. **Like death and taxes, security incidents are inevitable.**
   Security incidents are just a matter of time. Statistics support this. Across the three basic categories surveyed, the average company will experience between four and five security incidents a year—at least one of which will likely be a breach.

2. **Attacks by malicious employees are the most common.**
   Organizations often focus on external threats alone, but there is a very real internal threat that they may be facing—discontent employees who willingly leave the network wide open. This is why internal protection is also required to have effective cyber security. It's a combination of perimeter protection and internal protection that is most effective, making sure you never "rob Peter to pay Paul."

3. **Employee error is a problem that needs your full attention.**
   Did you know that employee errors account for more breaches than malicious actions and hackers combined? Many businesses do not realize this and that means they are more at risk than they realize against losses. Education/training is the primary solution, and a great backup is technical controls such as Data Leakage protection (DLp).

4. **The larger the enterprise, the greater the risk.**
   Organizations that have 1,000+ employees should expect to have approximately six security incidents per year—two of which will likely be data breaches. This is about 20% higher than smaller businesses. The best way to counteract this threat is to allocate the proper amount of resources to defending against it.

5. **Refine Return On Security Investment (ROSI) values with the Calculated Attack Frequency Figures (CAFF).**
   One of the primary variables in any ROSI calculation is the Annual Rate of Occurrence (ARO). Plugging these numbers into the ROSI

formula will yield more accurate figures than AROs derived at through guesswork. Due diligence is a must—always!

Acceptance of these five truths will help establish the proper business mindset to take action, which is what is laid out next. There are ten processes that will allow businesses to take serious action against cyber crime that is targeted at them.

TEN CYBER SECURITY PROCESSES

The goal is to stop cybercrime before it starts.

We've all heard that information is only valuable when it's heeded and applied. Through the ten cyber security processes that I'm going to share with you, you'll be a step closer toward application, which means that you are a step closer to data safety in your virtual world.

1. **Hold regular, consistent security awareness and training sessions.**
   Even with robust security controls, end users remain one of the weakest links in an organization. It's challenging to keep up with **Advanced Persistent Threats** (APTs) and unique attacks. To help circumvent this, training is required. It reinforces the behaviors that make employees more aware of threats or signs of an impending one. *Remember—end users can intentionally or unintentionally wreak havoc on your technology.*

2. **Develop and deploy security policies.**
   Despite everyone's best efforts, security breaches are inevitable— and, they are costly. Having standard procedures in place limits damage and exposure, which saves time and money. Having a comprehensive corporate security policy integrated into job descriptions and employee routines is a smart move, making it so security is not a low priority! *Remember—security policies are a living document, subject to regular reviews and updates to maintain their relevance.*

3. **Build a Security Governance and Management Plan.**
   It's far too common for security teams not to be exactly sure about which compliance requirements apply to them or even if they're meeting them. The result is a scramble when audit time comes. This is all solved by having direction from management that shows security is a priority and reporting on it is necessary.

After all, without security, the best of business goals cannot be achieved or maintained. Here are the three goals you should have when creating a Security Governance and Management Plan:

a) Recognize and prioritize your current security governance and management challenges.

b) Develop a security governance and management framework suitable for your needs at a minimal cost.

c) Implement your security governance and management framework using these tips and processes.

4. Optimize security operations without overspending.

A significant struggle for organizations who want to manage security risks and meet compliance requirements is their budget—which is usually minimal. Working smartly and effectively is the best way to get up-to-date and maintain that when it comes to both current and emerging security threats. Responding quickly with respect to volume, velocity, and the variety of security events that can take place is essential.

One challenge that many organizations face is the inability to demonstrate the value of security functions to leadership. The result is that security goals are not aligned with the overall enterprise-wide business goals or the various departmental and functional goals. Threats evolve every day and are often unforeseeable, as well as diverse. Because of this, protecting information is less about technology and more about contributing to the sustainability of your organization as a whole. It leaves many security decision makers asking, "What's next?" Here's what is recommended:

a) Make technology work for your people, not the other way around. Having well-defined processes can result in an operationally-effective Security Operations Function (SOF), which has a balance between security, cost, and ease of use.

b) Maximize your success and credibility by clearly defining your SOF mission, vision, and responsibilities.

c) Create an executable plan by assessing challenges, identifying gaps, and building an implementation roadmap.

d) Continuously improve plans by putting measures into place to gauge their effectiveness.

5. **Develop and implement a Security Incident Management Program.**

Accept that security incidents are inevitable; it's how an organization deals with them that will make or break them. Poor incident response negatively affects business practices, including workflow, revenue generation, and public image. To prevent this, a formal management plan is necessary. "Out-of-the-box" classifications shouldn't be relied on any longer, as they are too broad and easy to ignore.

Creating a streamlined process and formalizing a Security Incident Management Program will allow you to create a plan that is organization specific, therefore, more effective. This plan will allow you to more heavily monitor any incidents and analyze, track, and review them regularly. By doing this, you can lessen the likelihood of missing trends and patterns regarding the incidents that increase your chances of being re-victimized by the same vector.

Also, you want to establish communication processes and channels so they are in place well in advance of a crisis. *Remember, operating in a state of panic is neither efficient nor prudent when you are in a compromised security position.*

6. **Develop a Network Security Roadmap to lower incident costs and increase efficiency.**

Well-publicized breaches and compliance concerns are providing increased motivation for IT departments to begin focusing on forward-thinking plans, but they often have the misconception that it requires outside help, which means that unnecessary costs are accrued upfront. This can be averted either fully or to some extent by creating a roadmap, and it does not have to be overly complicated.

When you have the right people in place this type of project doesn't have to take a month or more to complete. In many cases, you can complete a roadmap in two days using information your organization already possesses. *Remember, many times when a consulting firm comes in, they will be asking you the same questions that you've already been asking yourself.*

7. **Secure critical systems and intellectual property against APTs.**

APTs are prevalent today and they will target any and all organizations that have some type of valuable intellectual property, or are laterally connected to a target organization. Companies must know how to protect themselves from these attacks, and also track and quantify an attack, respond to an attack, and ensure maintenance. A highly-used technique by attackers is a spear phishing attack—something designed to gain access to an organization's data through seemingly harmless email messages. Here are three steps that you can take to better defend against APTs:

a) Build the necessary multi-layered defense approach with specific tracking and monitoring capabilities so you can better defend, respond to, and investigate attacks.

b) Identify your risk by identifying the likelihood of an APT being carried out against your organization.

c) Ensure success by prioritizing your security gaps based on the importance and achievability of each measure.

8. Ensure cloud security environments are secure.
Hosted cloud environments, such as infrastructure as a service (IaaS) or platform as a service (PaaS), offer major IT and business benefits to organizations. However, security with the cloud environments does create numerous concerns regarding:

a) The ability for data privacy.

b) Confidentiality.

c) Integrity to be maintained in a cloud environment.

Even when you agree that you need to adopt a cloud service, it still remains challenging to evaluate vendors that have strong security offerings against those who do not. *Remember, it is also difficult to utilize security controls that are internally deployed in the cloud environment.*

9. Build security architecture and a roadmap for acquisitions.
When businesses stop being reactionary to cyber threats and become proactive, they will begin the process of staying ahead of, or at least on pace with, hackers. The solutions that have been set will guide you through the process of developing a formal plan that outlines the suggested security infrastructure acquisition policies for an organization. Remember:

Your plan should define appropriate security architecture and accompany it with an implementation roadmap, effectively defining the Cyber Security Process that an organization will use to build its security capability.

An enterprise that deploys security solutions according to a plan will have stronger security and spend less to achieve it.

The average cost, in time and dollars, to create a fully-customized security plan is eight months and over $100,000.00. Using the processes outlined here, you will be able to create a plan without those hefty expenditures.

10. **Build an Optimal Digital Asset Security Services Plan.**
The single greatest problem with digital assets is that they are usually unsecured, which allows for unwanted access and exfiltration of sensitive data. What is the result of this? It is usually a major loss of proprietary information; plus, major costs to an organization as they address detection, notification, data recovery, and legal expenses.

Limited budgets in organizations are the attackers' greatest joy. They know that a business that doesn't spend what is necessary to ensure protected data will be less secure across multiple areas, and that means that data is exposed. To be truly effective and optimize your security spending, a life cycle approach must be adopted that secures the entire life of the digital asset. *Remember, all assets need to be looked at together and across all touch points to create an asset security system dedicated to preventing data theft or leaks, while also focusing on doing the most with the least.*

YOUR CYBER SECURITY'S BOTTOM LINE

How you manage and address security is one of the most important decisions a business can make.

IT security threats are a complex process to understand, because it relies on people and processes being combined with the right products. Once we understand and acknowledge data security's importance, we can begin to implement processes that deter cybercrime and eliminate the most risks using the fewest resources.

About Raj

Raj Kosuri deeply believes in "Simple Living & High Thinking" and possesses extraordinary technical expertise in the field of Information Technology and business acumen. He wears a contagious warm smile on his face, which makes others comfortable with him even before he commences an interaction. Being a family man, he loves spending time with his family and ensures he makes quality time for them religiously. He waits to play his favorite soccer game with his nine-year-old son in his indoor make-shift stadium. He strictly likes to keep his official business countenance and his home disposition far away from each other.

Raj has an undying passion for driving from his youth, and is ready to take to wheels any time of the day. This unique mix in his personality has given him the right recipe to build up his IQ and appetite to run and grow business with the right spirit.

Raj believes that sharing knowledge and innovation are among the most effective ways of progressing – especially in the field of Information Technology. As a CEO & CTO he has authored *Best Practices for Business Rules Integration* in 2006 to share his expertise with the world. His chapter on *Mobile Application* is yet another way to do so.

Raj led his company to the "Deloitte Fast 50 & 500" four years in a row. He was awarded the honor of being named one of the Smart 100 CEOs by *SmartCEO Magazine* in 2009. In the same year, he introduced "Verdio" the Green PC – which won the NVTC Green Award for Small Business 2010. That was not all, in 2011, Raj released the Climetric Software designed to help Fortune 500 companies with carbon management and accounting.

Raj has very optimistically developed the **Green Data Center** in Danville, VA – spread across 27,000 sq. feet to offer cloud computing, disaster recovery and business continuity solutions in a virtualized environment. He plans to generate local employment there for those qualified – allowing them to stay in their hometown and work.

CHAPTER 17

IT REGULATORY COMPLIANCE

BY SEAN CONNERY

The most basic way to think about IT Regulatory Compliance is as a set of rules that ensure you are using correct procedures to manage data – based on your organization's industry. Sarbanes-Oxley (SOX) is a set of regulations that apply to publicly-traded U.S. companies. The Health Insurance Portability and Accountability Act (HIPPA) applies to organizations managing health data. Payment Card Industry Data Security Standards (PCI) sets regulations for credit card data. There are a plethora of other regulatory compliance acronyms that exist. They run the gamut of industries and organization types, including regulations such as the GCB (Gaming Control Board) for gaming casinos and the GLBA (Gramm–Leach–Bliley Act) for companies that extend lines of credit. This chapter will focus on SOX, HIPPA, and PCI.

COSTS OF SECURITY BREACHES AND STOLEN DATA

According to the Ponemon Institute's *2015 Cost of Data Breach Study: Global Analysis*, sponsored by IBM, the average cost of a computer breach to a company in the United States was $6.5 million. While closely following IT compliance regulations will not 100% guarantee protection from a security breach, it will drastically reduce your risk. Regulatory compliance is extremely important for many reasons, including defending against security breaches. Adhering closely to regulations can directly save your company money and provide peace of mind when managing health and credit card data.

If you are a consumer, you want to be assured your health records and financial information, like credit cards, are being protected by organizations following regulatory compliance as required. If a company or a medical provider you use was breached and your financial, medical health, and personal information was obtained by hackers, it could cause you great grief – even with a minimum identity theft.

The most common question I hear when discussing stolen data is, "Why do hackers bother doing it?" The answer is plain and simple; it is big business for cyber criminals. In 2015, credit card information was sold for a $1 per record, while medical records including a Social Security number averaged $363 dollars each. (Ponemon Institute, 2015)

Data breaches in the United States are most often the result of malicious or criminal attacks (49% of the time), with system glitches accounting for 28% and human error accounting for the remaining 23%. (Ponemon Institute, 2015)

CHANGING REGULATION AND AUDIT COSTS

The Sarbanes-Oxley Act of 2002, or SOX, was created to ensure corporate executives provide accurate financial reporting. It is enforced by the Securities and Exchange Commission (SEC) and resulted in the creation of the Public Company Accounting Oversight Board (PCAOB), a nonprofit corporation responsible for overseeing audits of public companies.

As financial reporting comes from databases, spreadsheets, ERPs, etc., IT has the responsibility of ensuring data accessibility, integrity, and accuracy. To comply with SOX, organizations must understand how the financial reporting process works and must be able to identify the areas where technology plays a critical part. The 2007 SOX guidance from PCAOB and SEC state that IT controls should only be a part of SOX 404 assessment, to the extent that specific financial risks are addressed, which significantly reduces the scope of IT controls required in the assessment.

In 2013, the Committee of Sponsoring Organizations of the Treadway Commission (COSO) was created. COSO outlines five aspects of internal controls: control environment, risk assessment, control activities, information and communication, and monitoring, which need to be in place to achieve financial reporting and disclosure objectives. According to the Protiviti 2015 Sarbanes-Oxley compliance survey, nearly 80 percent of organizations complying with Section 404(b) of SOX used COSO's new framework to guide their SOX documentation efforts in fiscal year 2014.

As such these new regulations are causing external and internal audit fees and labor to rise. Auditors become more stringent in applying audit tests, spending more time on the tests, as many of the areas subject to greater scrutiny relate to PCAOB inspection report findings. In fiscal year 2014, 58% of companies saw an increase in their fees; of those reporting, 40% saw fees increase over 15%. (Protiviti, 2015). Audit cost increases are occurring not only for external audits but also for internal audits. 54% of companies reported that the total time devoted to SOX compliance increased by over 15%. (Protiviti, 2015) During the initial one to two years of SOX compliance activities, and the year organizations transition from COSO 1992 framework to 2013 framework, organizations should expect to invest additional hours.

HIPAA, HITECH, AND ARRA

The Health Insurance Portability and Accountability Act (HIPAA) was enacted by the U.S. Congress in 1996. The American Recovery and Reinvestment Act (ARRA) was signed by President Barack Obama on February 17, 2009 and included the Health Information Technology for Economic and Clinical Health Act (HITECH), which intended to encourage the use of health information technology. ARRA also included a stimulus package which provided over $19 billion to support and promote the adoption of electronic health records (EHRs) for all Americans by 2014.

HITECH addresses the security concerns of transferring health information electronically and even strengthens enforcement of HIPPA. It added breach notification requirements to HIPPA and extends the reach of HIPAA to "business associates" of covered entities, a broad category which could include SaaS providers, IT services providers, lawyers, accountants, and more.

HITECH added fines for negligence, overseen by Health and Human Services (HHS) and Federal Trade Commission (FTC), and guidance for securing Protected Health Information (PHI). Electronic PHI remains unsecured unless it is rendered reasonably unusable, unreachable, or indecipherable to unauthorized individuals by more than one of the following:

- Data at rest has been encrypted as specified in NIST Special Publications 800-111 Guide to Storage Encryption Technologies for End User Devices.

- Data in motion complies with NIST Special Publication 800-52, Guidelines for the Selection and Use of Transport Layer Security (TLS) Implementations; 800-77, Guide to IPsec VPNs; or 800-113, Guide to SSL VPNs, or others which are Federal Information Processing Standards (FIPS) 140-2 validated.

PCI

Once a merchant is even suspected of a breach, their business is brought to an absolute standstill for a minimum of several days as a team of PCI certified forensics security examiners swoop in to investigate.

PCI affects anyone doing business with credit cards and payments such as issuing banks, payment processors, and merchants. PCI has global jurisdiction and dramatic consequences. Noncompliance consequences are dire. The Verizon 2015 PCI Compliance Report calculated exactly how dire some consequence can be: (Verizon, 2015)

- $3 to $10 per card for replacement costs (Verizon, 2015)
- $5000 to $50,000 in compliance fines (Verizon, 2015)
- Additional fines based on the actual fraudulent use of the cards
- Providing identity protection services

The bottom line is that the cost of data breach for a merchant – with less than 20,000 transactions – averages $36,000, and can be as high as $50,000. (Verizon, 2015)

STATE BREACH AND PRIVACY LAWS

As of January 12, 2015, all states and territories of the United States have breach laws except Alabama, New Mexico, and South Dakota. Two states, Massachusetts and Nevada, are leading the country in data protection through the passage and enforcement of two very specific laws.

Massachusetts enacted *201 CMR 17.00: Standards for the Protection of Personal Information of Citizens of the Commonwealth* in 2010. This law established minimum standards for safeguarding personal information in both paper and electronic form. Anyone that owns, licenses, stores, or maintains personal information about a Massachusetts resident must develop, implement, maintain, and monitor a comprehensive, written information security program applicable to records containing personal information.

The passage of this law led to specific reporting of data breaches, which provides an accurate picture of the scope of the problem in Massachusetts. Overall, the results of this reporting strongly illustrate a greater need for security measures that focus on intentional wrongdoing and internal protocols.

Nevada enacted *Senate Bill 227*, which forbids a data collector from transferring "any personal information though an electronic, non-voice transmission other than a facsimile to a person outside of the secure system of the data collector, unless the data collector uses encryption to secure the security of the electronic transmission." It also forbids Nevada data collectors from moving "any data storage device containing personal information beyond the logical or physical controls of the data collector, or its data storage contractor, unless the data collector uses encryption to ensure the security of the information."

These provisions put considerable burden on business to control its employees use of flash drives, notebooks, tables, and other devices and media that can easily be loaded with sensitive data and removed from the employer's premises. Under the data security breach notification laws already in effect in most states, losses of unencrypted notebooks and other devices that contain personal information can obligate an organization to notify affected persons. The Nevada law turns the legal screw even tighter by making the act of removing the unencrypted device from the employer's premises a violation of law in itself, even if no security breach results.

STRATEGIES FOR SUCCESSFUL AUDITS

When your business is subject to IT regulatory compliance, audits become a very important part of that business. There are several strategies to help the audit process go smoothly and end successfully.

To pass or not to pass a portion of an audit, that is a tactic. As many of you may question why you would fail a portion of an audit I will start there. I worked at a large publicly-traded company and at the beginning of the year I had an approved Opex and Capex budget. In the Capex, there was a complete backup solution as we had hundreds of servers all around the world and we wanted to make certain all data was being backed up and that we could quickly restore services if there was an issue. After beating up vendors and selecting the solution that was best, I went to issue a purchase order. Even though it was an approved expenditure I still could not get a purchase order for this project.

Although we passed the audit with flying colors, attention was brought to auditors who reported we needed a better backup solution. The purchase order was then funded and approved within a week. Sometimes there is value in bringing attention to an issue for the good of the company. It can motivate and convince them to move forward. That said, we are not advocating failing an entire audit, which most likely would be a resume-generating event.

Another strategy for passing an audit is to first outsource to an auditing company to conduct a pre-audit. This provides an opportunity to secure any holes about origination funds. You can also leverage a pre-audit to defend the actual audit, if need be.

Documentation and conducting a pre-audit ensures your environment is in alignment, which is very important to a smooth audit process. What is actually deployed, and what is in your documentation, must match. Auditors are confirming alignment of your policy and what is implemented in your environment. If your policy says you are checking for new Anti-Virus definitions every four hours and you check the software and it is set to check every hour, even though you are exceeding policy you will still fail as you are not in alignment.

A way to remedy this in a SOX audit was to have policy documentation regarding anti-virus, include specific verbiage on if employees were to disable or tamper with, the levels of discipline; but when it came to settings on how often we updated definitions, our documentation stated that what was configured in the Anti-Virus management console was our policy. In this case, the auditor took my policy and a screenshot of our settings in the management console, and they matched.

For successful implementation of regulatory compliance, and the corresponding audit, you must obtain understanding, guidance, and acceptance from senior leadership. In Certified Information Systems Security Professional (CISSP) training, it is often discussed that IT are data custodians implementing the data at the owner's request. This is an important to remember when preparing for an audit.

Determining acceptable risks and associated costs is another part of preparing for a successful audit. If you are subject to multiple regulatory compliances, generate a matrix of each category of compliance. This matrix of "rules" should take each regulatory compliance and record it down into each rule, such as backups, password policy, use of remote access, etc., then correlate that with each regulatory agency.

For example, perhaps one regulation only stipulates that you have backups on-site in a fireproof safe. The requirements of a second regulatory policy state backups must be off-site by twenty miles. By developing a matrix, you can look at all the regulations and choose to follow the maximum rule in order to streamline compliance.

OUTSOURCING AUDITS

For the last ten plus years CIOs have been operating within the mantra to "do more with less." There is a proliferation of data and yet resources for operational budgets are stagnant, or even worse, shrinking. The IT component of the financial audit can be as simple as a member of the IT team sitting with an auditor for two to four hours to provide necessary information, to a SOX audit that can take three to five weeks. It can be difficult to surrender a senior IT security employee for that amount of time. For that reason alone, outsourcing your compliance audit can be a worthwhile investment.

There is also great value in hiring someone who has done something numerous times, as opposed to internal staff who has never done an audit, or at best, have done audits infrequently. Do you want to be responsible for managing a complicated audit, with very real and potentially-damaging consequences if failed, or would you rather outsource to someone with the time and expertise to do it correctly?

Works Cited

Ponemon Institute. (2015). *Cost of Data Breach Study: Global Analysis.* Sponsored by IBM.

Protiviti. (2015). *Sarbanes-Oxley Compliance Survey.*

Verizon. (2015). *PCI Compliance Report.*

About Sean

Sean Connery is President and co-founder of **Orbis Solutions**. He is a Certified Information Systems Security Professional (CISSP), Microsoft Certified Systems Engineer (MCSE), ITIL certified (Information Technology Infrastructure Library), VMware Certified Professional (VCP) and has been nominated twice for Vegas.com's Top Technology Executive of the Year. Not to mention the numerous other certifications with Microsoft, Symantec, VMware, etc.

Sean's 30 plus year career includes being a senior executive in Hotel & Casino, Casino Gaming, a national jewelry chain, and software development ventricles. At the end, Sean and his team provide Information Technology to business. What he really loves is working with clients and to engrain himself in their operations to lavage technology as a competitive advantage for them. He will involve himself with a motor cycle dealer for hours, then move on to a medical operation, then a CPA firm or a Hotel Casino to name a few.

In addition to working with his clients, he has been a speaker at the Microsoft store, CEO/CFO, Nevada accountant groups on mobile device security, Cyber security, and regulatory updates.

Outside of computers, Sean is a member of TPC and plays 18 holes in two hours and fifteen minutes Friday mornings. He is a wine maker of Bond Bordeaux – an award-winning wine, both a Napa and a Chilean version. Sean likes to learn everything about what he is doing, so he took scuba lessons 15 years ago and kept learning and became a Dive Master, and kept learning and is now a PADI Master Scuba Diver Trainer (Scuba instructor).

CHAPTER 18

CLOUD STORAGE:
PRIVACY, SECURITY AND SERVICE CONSIDERATIONS

BY VIJAY NYAYAPATI

August 31, 2014 - A collection of private pictures of various celebrities containing nudity released online in public forums. The photos were being passed around privately before their public release. A few weeks later, thousands more of nude images belonging to various celebrities and individuals released by hackers.

September 11, 2014 – Nearly 5 million Google passwords leaked on a Russian site.

October 17, 2014 – Nearly 7 million Dropbox passwords leaked online apparently via third Party services.

The list goes on. A majority of the incidents are not reported or discovered.

The emergence of cloud services has provided a significant growth in offerings to the IT industry and very importantly, to users in general. This is an emerging technology that has taken computing by storm, especially storage available at lowered costs, and in many cases free for a limited capacity. It gives users an unmatched capability of highly available backups and archives, as a managed service – that too with an offsite option. Without a doubt, one of the primary uses of cloud computing is for data storage.

Cloud storage introduces several useful features to users – such as easy accessibility, high availability, multi-device synchronisation and collaboration. Whilst all these features provide unprecedented value-added to users, they come with one significant issue: data privacy. Amongst one of most dangerous incidents of privacy breach concerns is identity theft – where user data is accessed without consent and, in extreme cases, used for activities like extortion and blackmail (as in the case with celebrities in the August 2014 incident).

PRIVACY

Cloud Storage & Privacy Issues

Many solutions now exist, such as Dropbox, Box, SpiderOak, GoogleDrive and OneDrive. Some of these offerings have been conveniently facilitated by email service providers as extensions to their core email service to integrated storage services - such as GoogleDrive from Google and OneDrive from Microsoft. The need or the desire to access data anywhere at any given time and a rise in mobile device usage has necessitated that data can be accessed from anywhere at all times.

Privacy is of importance to everyone, but one of the drawbacks of cloud storage is that it can accidently disclose information in unforeseen circumstances. As a basis for cloud data privacy, an undertaking of trust is necessary between a provider and a consumer to safeguard user data. Just as people implicitly trust banks to safeguard their money and doctors to cure their illnesses, they, without any consideration, trust service providers for safety of their data the moment they choose to use the provider services. Legal clauses and technical mechanisms to do this follow afterwards, but this basic acquiescence of trust is established the moment cloud storage provider selection happens. Then, formal acceptance of trust is enacted when "use policy" or "terms and conditions" is accepted by users at the initiation of service usage.

The problem, most times, is that at this initiation point, where users are in fact informed of what type of data is being collected about them and how it will be used, they unassumingly accept. Only some users take the time to read privacy policies, but still lack the understanding of what the risks are even when they do. As an example, in 2012, 98% of Dropbox users were using the free services offering and therefore were potentially obliging for data mining activities by Dropbox advertising sponsors.

However, all free cloud storage services have some level of security weakness that could result in data leakage without user awareness. The reality is that it is difficult to refuse free services, especially when several gigabytes of free data, with backups, is made available to users by simply signing up. These come at the risk of privacy issues.

Classifying Privacy Issues

We can begin by asking what authentication mechanisms are in place to ensure the connection to cloud providers is secure. Then, once a connection is established, how do we know the data being transferred across to the provider is not being snooped upon? Finally, when data is eventually residing on the storage providers' site, what guarantee do users have that providers do not accidentally make it available to other cloud users?

Data at rest and data in transit are major issues of cloud storage. To compound the problem, online collaboration now allows for multi-edit capability of common documents, using technology like Google Docs, over shared storage platforms such as GoogleDrive. As part of this collaboration, what happens if a URL-based shared document lands in the inbox of an unintended recipient? Can there be co-existence of collaboration whilst maintaining privacy?

To mitigate privacy issues, there is a need to understand how data is secured by the providers (data at rest), how data is secured whilst getting to the provider (data in transit) and how data is being secured when sharing with others (data in collaboration).

To users, characteristics like availability and reliability also matter, along with acceptance of legal terms and trusting providers to safeguard data. Cloud regulatory issues are a major challenge in cloud adoption. For example, if the cloud storage provider goes bust, what happens to the data they hold and who then owns it? Users expect cloud storage services to always be available, but statistics of cloud storage performance indicate that users are bound to experience some loss of data during some future cloud event.

SECURITY

Data in Transit

Data in transit can be described as active data which is on its way from source to destination – the data passes through several intermediate points over the Internet before the download or upload actions are completed.

The data in transit is prone to spoofing. The lack of a feasible solution in the mobile cloud storage security is an obvious loophole in ubiquity data privacy. Mobile devices, where the security level is factually weak, is in fact the point from which attackers intrude into cloud storage systems. Provisioning of security services (encryption and decryption) can reduce this risk.

Data at Rest

User data residing on cloud storage servers that can be read or updated can be referred to as data at rest. When this state is achieved (basically upon completion of an upload task) it is assumed data is safe.

On June 20, 2011, 45 million Dropbox user accounts could be accessed without passwords for four hours, by simply typing an account email address. No doubt, this is a significant area of security concern in cloud storage adoption. From the moment a computer is connected to the Internet, data is being collected for data mining. Is the data being used for advertising and marketing purposes? That is, is it being assessed by providers for any other purposes than providing storage capability? There is also a serious security flaw in Web services-enabled cloud access-control where users enable a service such as Facebook to access data managed by other services such as Google, and therefore providing the former authentication credentials for the latter, effectively giving the first service full access to the second account.

The most logical mechanism to stop unwanted prying on this data is in ensuring even if the data is accessed, it is in an unreadable format. The best way to achieve this is to use third party solutions that allow users to encrypt files locally before they are uploaded to cloud storage services. These third party tools work well with common providers like Dropbox, and are seamless to users, whilst automatically securing data with industry-standard security. However, widespread user awareness of such encryption tools could be limited in spite of many products such as Boxcryptor and Viivo being available free of charge for personal use.

Data in Collaboration

Offerings from vendors such as Google have made it possible for multiple users to collaborate across the Internet – i.e., facilitating multiple users to manage and edit common documents stored on cloud storage systems at the same time. This can be referred to as data in collaboration.

Not many people may use private sharing or public sharing types of cloud collaboration – instead they prefer the URL sharing type of cloud collaboration even though it is notoriously dangerous in that anyone in possession of the URL can effectively access the data. As an example of this, if Joe shares a URL link with Maggie, then user Maggie can re-share the link with John – problem is Joe never intended to share documents with John in the first place. There can also be an accidental recipient of a shared link, based on an erroneous email address by the originator.

A method known as Rights Management provided by authoring tools can give users a mechanism to protect intellectual property

SERVICE

Legality

It is reasonable to assume most users pay little attention to the many pages of "Terms & Conditions" when using services online. Users want to "accept" T&Cs and just get on with it. The legal issues that could arise in cloud services are wide ranging, especially with those regarding privacy of data and data security, as they relate to protecting personally identifiable information of individuals, and sensitive and potentially confidential business information.

The primary difficulty many providers such as Google and Microsoft face is that they have a single privacy policy for all of their services. The limitation with this is the general vagueness of clauses to cover all of their products. Therefore, the right to access and deal with user content is usually not explicitly stated. This gets worse with providers such as Apple who, for example, have a clause in censorship and "the right to delete" without prior notification any content it finds "objectionable." Primarily owning data and then having the responsibility as its custodian, have been noted as fundamental different aspects to legality. Users can own data as the originators. However, providers may become responsible for this role too because where the data actually resides and what laws it may breach, are related concerns.

In the case with file-sharing site Megaupload, where the site was taken down for movie copyright violations, based on uploads by some users that were contrary to U.S. legislation – what this also meant was that every other global user with data on this site also permanently lost their data, irrespective of their individual involvement with the matter. This also touches on the boundaries of copyright. Once users upload content on providers' sites, who owns the content after the upload?

There is no easy way to solve the legality issues. In fact, it remains one of the most unanswered aspects of cloud computing in general. Without exception, legal clauses and restrictions apply to every cloud storage site. By using the services, users accept these legal clauses. However, the necessity for investigation and comprehension on what these mean for them and knowing the rights and scenarios under the legality clauses cannot be understated.

Reliability

How can users be certain that services being used will always be available and therefore reliable? As an argument, if the integration to underlying storage (such as GoogleDrive) works via an email account (Gmail), how vulnerable is the continuity of that service, for example, when the email account is temporarily suspended by Google for suspicious activity? This is not uncommon if someone other than the account holder tries to access the account several times, unsuccessfully. This then means the data in GoogleDrive is no longer likely to be accessible by the account holder.

Another scenario could be to imagine what happens when a provider site is down for extended periods of time without warning, and when it is operational again, some data for the users is missing? Personal data is by no means as safe as users thought. As soon as reliability-related events take place, users lose control over their data as they are unable to control data loss. Reliability issues such as denial of service attacks and social networking attacks can also prevent access to data.

ENCRYPTION

Various surveys conducted of user concerns in cloud storage found security was the primary concern. Not surprising! However, it is possible to secure storage services by encrypting data relatively easily using sophisticated cryptographic mechanisms, but all very seamless to the user.

Providers such as SpiderOak and Mega have a default policy to encrypt data and handing the key to users. This also means the vendor cannot decrypt data and the user has to be notified by default. Other vendors encrypt data, but retain the ability to decrypt it. This represents a service where the provider encrypts the data. Although this "server-based encryption" is an option, the data may still remain vulnerable since server-based decryption keys must be accessible to the storage website.

Within the security dimension, three characteristics were identified that needed attention. The client and server-based encryption appears to solve all of them. If data is encrypted before it is transferred, then it is unreadable to a rogue party whilst in transit or by the provider while at rest.

Despite this, the third – data in collaboration – is somewhat trickier. It gets cumbersome because encryption keys need to be shared with all collaborators who require access to data. This means an encryption key distribution mechanism involving multiple people must exist. This then poses another challenge – how are the users to know that the authorised collaborators have not further redistributed those encryption keys, unknown to the originators? Therefore, these application-level cryptographics may in fact provide a false sense of security to users, and the key to ascertain their effectiveness is therefore in examining, not just assessing, what is encrypted – but also how the decryption keys are managed.

CONCLUSION

Encryption can prevent unauthorised access to user data, and further secure the session traffic from snooping. It is surprising that not all cloud storage providers encrypt data in transit as a default option. Since most applications cannot process encrypted data, it is ideal if data is secured before arrival in the cloud. Encryption at the client end means the provider does not know the key, so this not only addresses the data at rest issue but further secures data in transit.

Irrespective of policies, cloud storage providers must commit to fundamentally protect the privacy of user data. The user must then decide if the provider can be trusted. Trust is the basis for security and privacy in digital systems and it is inherent in data privacy through cloud-computing services. By using cloud services, users initiate a high level of trust between themselves and providers.

The choice to utilise benefits of cloud-based services is ultimately one made by users. A reasonable mitigation here is to sign up for services that offer acceptable levels of encryption, uptime and availability. Whilst this may be true for business users, personal storage users whose services are generally acquired for free, may struggle to get such guarantees. Irrespective of the type of users, prior to selecting which service provider to use, a useful first step could be to evaluate performance of the providers' security, reliability and availability over an extended time period.

About Vijay

Vijay Nyayapati is the founder of **Computer Support Company Ltd.**, an IT company based in Auckland, New Zealand. Vijay brings over 22 years' experience in the IT industry.

Vijay has been working with technology since the 80s when his dad bought him his fist 80286 computer that he had to share with his brother, during the MS DOS 3.x era. This naturally got him interested in pursuing an education and career in information technology.

After having worked in the industry for a few years and seeing that many small businesses were being overcharged for technology services whilst not getting the promised level of service, Vijay formed his company in 2004, with a different strategy to most IT providers. The business model used by CSC does not permit profit generated from capitalising on client technology problems. Instead, networks, desktops and servers are managed proactively to minimize downtime and save money for clients.

With the constantly changing security scenario and an overload of information available to small and medium-sized businesses, Vijay realised that SMBs did not protect their IT systems and networks well enough to prevent attacks and data theft. The security focus by many other IT providers was reserved for the 'elite' corporate clients. Therefore, he has made it his personal mission now toensure that SMBs protect themselves against this growing threat, without breaking the bank!

For more information about Computer Support Company Ltd. or to learn how you can protect your valuable data, visit: www.csc.co.nz or email Vijay at: vijay@csc.co.nz

CHAPTER 19

SINGLE SIGN ON AND THE CLOUD

BY MATTHEW A. KATZER

The cloud is changing the way our businesses operate. Owners, employees, and contractors can work anywhere in the world, at any time, and use any device to access our business resources. The productivity that results from this type of freedom is welcomed, but it does create a new problem for business owners—managing these resources efficiently.

A STEP BACK IN TIME

In the pre-cloud days, computer security professionals worried about physical structures and described computer security as a layer of boundaries, which were known as COMSEC Boundaries (Computer Security Boundaries). The US government used this broad definition, and established a classification of access with levels of security clearance and used computer Security Boundaries to control access. As an example, a building would be classified as an "A level building" and you would have a room in the building with a "B level" classification. To enter the building, you would need clearance at the "A level" and to enter the "B level" room you would need to have a security classification at the "B level." It was a very manual process and some facilities used guards. They would stand there, checking to make sure you were at the correct security level to enter a room.

As businesses began moving more services to the cloud, a new problem was recognized: the limited ability to control access to third party services. We had to start controlling access to our data and grant individuals permission to use the services we subscribed to. The days of the building guard controlling access were over! In our secured building example, if an employee who had access to the "B room" was fired, all we had to do was tell the guard they no longer had access to it. Problem solved! In the cloud world, we control access by user IDs and passwords. In the fired employee example, we can block the user's access to the business application and data. This works great, but what happens if we have thousands (see Figure 1) of different applications, and multiple users? That creates a challenge regarding what we do next.

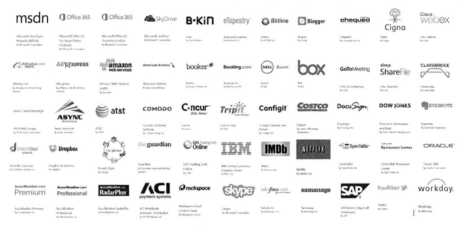

Figure 1. Sample of Different Cloud Identities (Courtesy of Microsoft)

WHY IS THERE A DEMAND FOR SINGLE USER IDENTITY?

In our earlier example, it was easy to make changes for a few people, but extremely complex if changes were needed for hundreds or even thousands of people. Expand this even further, where each user can access ten different cloud services. This is where Single User Identity comes in handy. A Single User Identity is a single user sign on that is used to access all cloud services.

Single User Identity is simple to set up. You configure your primary cloud services to treat that service as a master and link secondary services to the account. When you change the user password in the master service, all related services passwords are changed. Currently, this capability is shared by multiple different services, including Ping, Okta, Office 365, and Google Apps.

Today's business environment is driving the necessity to have Single User Identity. There is no more effective way to control access and information flow. Single User Identity allows additional security login enhancement, such as two factor authentication. Two factor authentication links the cloud services login to a physical device—such as your phone.

LOOKING AT OUR EMPLOYEES AND THEIR USE OF OUR BUSINESS SERVICES

If we step back and look at the employees in our businesses, we see the changes in the business environment with the cloud. We no longer have the capability to control the "building," because employees are bringing their own devices to work, running business applications, and using multiple different cloud services to manage their work (and personal) environment (see Figure 2). Plus, they can work anywhere in the world. All of this has resulted in our IT management becoming more complex, because the employees are now mixing their personal Identities and equipment into our business environments.

Figure 2. Different ways Businesses use IT services in today's Cloud environment (Courtesy of Microsoft)

As an employer, we now need to incorporate the necessary controls on the employee's personal equipment (and users accounts) so we can allow the employee to perform the required work. One of the issues in today's cloud environment is Business Intellectual Property. How do we business owners manage our business assets with employees that work from anywhere and at any time? The common thread in this scenario is the control and management of employees through a Single User Identity. With this approach, it no longer matters which "identity" you use as the cornerstone. (Do you use a Google Apps ID or a Microsoft Office 365 Organization ID?) It is more important that you pick one and use that as the standard in your organization.

An organization's Single User Identity should be derived from its core cloud or on-premises security server. By doing this, the organization's Single User Identity becomes the cornerstone of the business. Businesses are at the juncture where they must replace the building security with a better security—the Single User Identity that is controlled by the business.

As a business grows and expands capabilities, those capabilities become more apparent, because a Single User Identity can now provide a business with control over the following:

- User being able to bring any device into the organization
- User business identity can be granted to third party cloud services
- User can only use those company applications that are approved
- User can only access company data that has been approved
- User can access data anywhere that is approved by the company
- User security and data access is managed by the business through a common identity

We begin delivering the control and security of business information to the same level that we had with our old secured building. Single User Identities allow you to access data and services stored in any cloud solution; it also leverages the local security identities (often referred to as domain or active directory credentials) and works across any device. If you make things easier for your employees (with appropriate security governance issues fully addressed), the employees will perform to their potential and the business will grow.

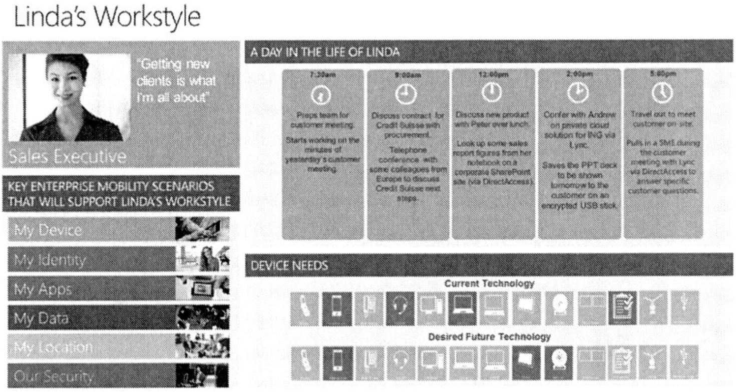

Linda's Workstyle

Figure 3. Typical User Business Life mapped to the Single Identity (Courtesy of Microsoft)

In Figure 3, we see a typical user that has different business needs, and requires access to different levels of services. We need to provide this user with the tools to access the different services needed to complete her job. This is where single identities are used to manage the employee work environment. This idea, while not new, has not been applied to the public cloud. Large enterprises always had the capability to manage the resources inside a secured building. What is different today is that businesses (of any size) need to manage these resources for their organization. As business owners, we need to manage the cloud services for which we have subscriptions, and use Single User Identity. The benefit is that Single User Identity becomes a powerful management tool for our business.

HOW DOES SINGLE IDENTITY WORK?

Single Identity is simply the linking of our on-site and 3rd party services into cloud services (such as Microsoft Office 365), and using the cloud services as the information broker to coordinate with other external business services. How businesses accomplish this depends on the business size. Larger enterprises will have the resources to manage these activities from their own data centers. Smaller businesses will manage this activity from third party services such as Microsoft Office 365 and Azure User Identity services. In Figure 4, we are viewing what happens in business when we link on-site security servicers (known as Active Directory) to Microsoft Azure services to access third party software applications, and how, when we link these resources, we enable a new level of security.

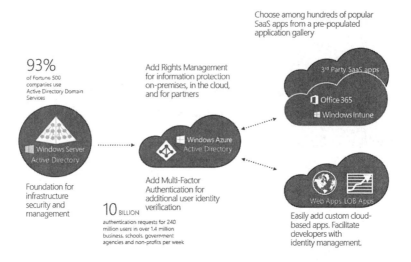

Figure 4 Linking Onsite and Cloud services for Single Identity Access (Courtesy of Microsoft)

The first step in the deployment of Single User Identity is to look at what we are trying to accomplish. For example, a company might use the following business applications: Dropbox, QuickBooks Online, Twitter, and maybe Pinterest. These services are building security for their own use and there is no convenient way to manage these applications. As an example, suppose you have developed a social media presence in the market, and a key employee leaves the company. How do you manage the user account of the employee who left? What about their Dropbox account or Google drive data? It is a "hiccup," maybe, with one employee. Now expand this across many employees and you can easily see the problem. In most cases, you have a friendly employee who will turn over the information to you and you can change the passwords. However, in some cases, you will need to use legal measures to get access to business data, because the ex-employee was using a personal account. When we use single user Identity, we are keeping secret the passwords for the third party services, and just granting access to the user to use those services. When an employee leaves, we remove their access and maintain control over the business resource. As business owners, we must control information in our business for our businesses to survive and grow.

WHY SINGLE USER IDENTITY VERSUS DEVICE MANAGEMENT?

When we look at Single User Identity, we are looking at the resources we use in our communications to external parties and how we manage the activities. We have to step back and look at the basic structure that the single identity is built upon. Too often we are trying to place a security strategy around a device, when the security must be built around the user. This is the fundamental switch that we see in today's business environment (see Figure 5), a switch from a device-centric model to a user-centric model. This is what we mean by Single User Identity: The ability to focus on the business' most important assets, which are our business data and our people.

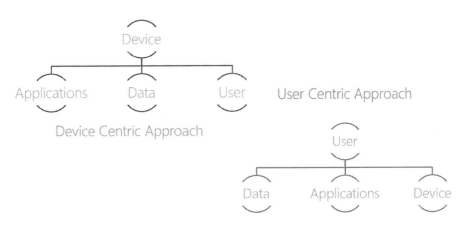

Figure 5. Single Identity being changed from a device-centric model to user-centric model (Courtesy of Microsoft)

Single User Identity is nothing more than managing those assets under a common identity (usually an email address). This user-centric approach is what allows for business growth. Adding additional cloud applications is easy, and this can be completed by the IT Manager/business owner. In Figure 6, we see that we can selectively add a link to the third party application. The application is linked to Office 365, and the user who has accounts in the third party services has immediate access. In this case, we are adding Google Docs to be managed by Microsoft Office 365.

Figure 6. Using Single Identity to centralized user information under one Business ID

Microsoft Office 365 Services provides us with tools that we use to determine the types and number of business cloud applications. Or, we can add them directly from our gallery of cloud applications. In this example, we added a DocuSign business account and linked the account to our Office 365 account. We can selectively enable Office 365 users to access the business' DocuSign account, without providing the user the DocuSign's password. The users now access DocuSign (via the myapps. microsoft.com portal) using the Office 365 account as a Single User Identity sign-in. We can also link a Google Apps account to Microsoft Office 365. In all of these examples, we are linking business assets to the users to improve productivity and maintain security. As a business owner, you can selectively provide your employees access to third party services based on the business needs (See Figure 7).

Accessing Company Information?

- ▶ Classify data according to sensitivity and business impact
- ▶ Differentiate access to data based on identity & role
- ▶ Offer an "Opt-in" model for readying devices to access corporate resources
- ▶ Deploy robust firewall and perimeter protection; maintain an 'assume breach' model

Changing perspective: From device control to data governance

Figure 7. Limiting access in using Single Identity to manage data Governance (Courtesy of Microsoft)

HOW CAN SINGLE USER IDENTITY HELP ME MANAGE MY IT RESOURCES?

One of the most overlooked aspects of Single User Identity is the consumption of IT resources. Traditionally, Single User Identity has not been used because of the amount of overhead associated with using it. This perception usually takes place when user passwords are not linked to a cloud service. Cloud services such as Microsoft Azure and Office 365 makes this linking easy and transparent to the users.

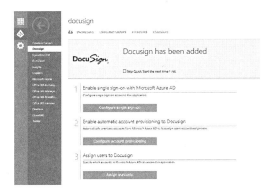

Figure 8. Assignment of DocuSign service to a Single Identity Office 365 Account

179

In Figure 8, we have linked the common applications to a company's Office 365 cloud account. For example, we are now linking the Microsoft Office 365 services to DocuSign. This is the way users can access applications with a common identity. Once you develop a Single User Identity (or Cloud Identity), there are additional services that you can make available to your users.

In reference to Figure 9, we have extended the capabilities of Single User Identity, by allowing users to have local control over the password reset (and access to additional applications in our inventory). The user can register to self-manage password reset, and the process will require the user to register their cellphone. Microsoft's self-managed password reset is a two factor approach. Two factor password reset is built around something that you have (your cell phone) and something that you know (your email address or phone number). The combination provides a very strong approach to allow the user to self-manage password resets.

Figure 9. User control over password reset and Single Identity

Self-managed cloud identities also allow the user to control access. The user can request access to groups and applications on demand. This leads to more seamless management of user permissions and service permissions without the direct knowledge of the user accounts. By doing this, managers and business owners gain the ability to control information and access for the employees in the organization.

ONE LAST LOOK AT SINGLE USER IDENTITY

We have delved into why Single User Identity is important and how this leverages the different third party applications that are available to businesses. Business owners that use a Single User Identity service have the ability to manage all resources that the business uses, which is a significant factor in both management of the business, and its growth. Single user identities can easily be transformed to a Cloud Identity, which uses the same security structure, but with no on-premises resources to manage thousands of third party services. In Figure 10, we have a Cloud Identity that is used to control access to documents.

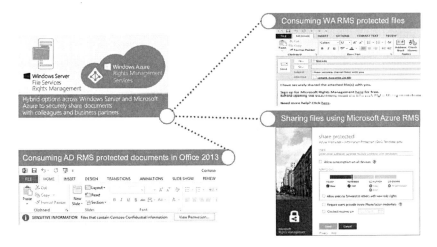

Figure10. Rights management with a Single Identity or Cloud Identity (Courtesy of Microsoft)

SINGLE USER IDENTITY TODAY PREPARES BUSINESSES FOR TOMORROW'S GROWTH

The examples in this chapter are based on using Microsoft Office 365, and Azure Active Directory management services. There are other services that you can use, but my professional experience has proven that the Microsoft approach is more seamless for the end user. It is also effective for companies from 1 to 100,000+ employees. Is it time for you to reduce your business risk and optimize your business's performance and growth potential? Then it is time for you to implement Single User Identity.

About Matthew

Matthew A. Katzer is the President of **KAMIND IT**, a Microsoft Gold Partner, and author of the best-selling cloud book: *Office 365: Managing and Migrating Your Business in the Cloud*. He is currently the President of the local chapter of IAMCP (International Association of Microsoft Channel Partners) and active in local business communities.

Matt's focus with cloud solutions started in 2009, as he was looking at ways that businesses could grow quickly and with reduced operating costs. Matt holds a BSEE from the University of Michigan and an Executive MBA from the University of Oregon. Matt's greatest satisfaction comes from helping his customers become competitive in an increasingly technology-driven world.

CHAPTER 20

SOCIAL ENGINEERING: THE BREACH HIDDEN BEHIND A FRIENDLY SMILE

BY DANIEL P. MARCELLUS

We've all received emails similar to this:

Dan,

We believe that your account has been put at risk of being compromised. We've taken preventative actions and assure you that you are protected. If you could you please verify your correct address and Social Security number for our files, it would be appreciated. We don't want your services through us to be disrupted in any way. We're sorry for the inconvenience, but assure you that we are taking every step possible to ensure your information is secure.

STOP! The sender of this email isn't looking out for you; they are looking out for their own interests and what they'll appreciate most is you sharing your personal information with them through a "reply." What happens next? They will have opportunities to sell it and earn some money, or else become you for long enough to enjoy a spending spree at your expense. It can be tough to accept, but behind those caring words and virtual smiles, are often malicious intents.

Any reputable business that you conduct business with will not ask you to email personal information—ever. If you received an email, they'd ask you to call them, which would give you the opportunity to make sure you were calling the right place. Or they would call you directly. The practice of virtually sending these types of messages is called social engineering.

By definition, social engineering is the art of tricking someone into providing access to something they should not have access to via social interaction and/or impersonating someone else.

Social engineering is one of the greatest threats that any organization encounters today. Its sole purpose is to access equipment and physical or electronic data to retrieve data without permission. *In these situations, cybercriminals are taking advantage of general human acts of kindness to commit a crime.* And since it's hard to always tell who is legitimate, it is important to understand the full scope of social engineering. Businesses try to protect against this, but it's a combination of consumer awareness and business efforts that will be the most effective.

PROTECTED HEALTH INFORMATION: A GOLD MINE FOR SOCIAL ENGINEERING

When most individuals think of Protected Health Information (PHI), they think of their basic medical history—if they have allergies, previous surgeries, or any chronic conditions they are treated for. When a cybercriminal thinks of your PHI, they are digging a little deeper. They are focused on all the personal information you provide, including:

- Full name and birth date
- Social Security number and insurance information
- Address and employer information
- And, all the details to fill in the gaps and create a complete profile that is mirrored in an unknowing individual's image

The big question is: *Whose job is it to protect PHI?* The answer is that it's the responsibility of those who work in the medical field, whether as a service provider or in a support role, such as claims processing and insurance companies. The guidelines for consumer's information being protected exist in the Health Insurance Portability and Accountability Act—HIPAA. All entities that have access to PHI in some capacity to fulfill their business role must follow the HIPAA guidelines, which are to take every reasonable precaution to protect a consumer's private information from being accessed by an attacker.

When it comes to social engineering, attackers are specifically looking for: PHI printouts, databases, passwords, knowledge of physical location of PHI. *This information is their gold mine, which is why organizations that work in healthcare are big targets.* To make sure that medical organizations take their best efforts to prevent breaches, HIPAA put some very serious consequences in place that will be enforced on businesses that don't take the act seriously. **Breaches equal massive fines from $100 up to $50,000 per record violation, depending on the severity of the breach**. On average, your typical medical record database holds tens of thousands of patient records, all of which contain PHI.

THREE TYPES OF SOCIAL ENGINEERING ATTACKS

Although cyber attacks indicate virtual theft, there are three crafty ways that an attacker can pursue and retrieve data that is not theirs. And seemingly "insignificant" data also has value, which means businesses are taking a great risk by assuming their data does not matter, because it does.

1. *In person*: It's not often that we think of someone stealing confidential information in person, but it can happen. This strategy can be as simple as that guy casually smoking by the entrance to your work building that "rides your coattails" to gain access at a secure door. It seems like it comes from a spy novel, but it is a legitimate and often-used attacker strategy.

2. *Remotely*: When an attack is executed remotely, the attacker can pretend to be someone else, such as a friend, boss, or co-worker, but they are asking for a bit of information that will be valuable to them for some reason. This is called phishing, and there are far too many people that take the bait. For a business, it means the potential for stolen data and malware infecting their system, too.

3. *Hybrid*: A hybrid attack is well thought out and deliberate. It is a way to infiltrate a business environment, build trust, and inadvertently have a business accept the attack methodology. An example would be a business' vendor who has access to information to perform their service and the business believes they are doing so in good faith, but in reality, they are taking data illegally and using it for some sort of gain.

Businesses that become vulnerable to any of the three types of social engineering attacks are in great jeopardy. Once a breach is uncovered, the chances of that business remaining solvent are significantly reduced, if not eliminated, depending on size.

WHO INVITES CRIMINALS IN?

Most every time a social engineering plan commences, the one doing the attacking was invited into that business without the business even realizing it. It is incredibly easy for a business to become a target due to the "invitation," and no amount of technology will make you 100% protected.

Employees
Employees are notorious for allowing physical access to attackers, letting an unknown person into a business to ride their coattails. Someone may be standing outside that door holding a heavy printer and it looks like they're just waiting for someone to let them in. The employee is showing up at work and wants to be nice so opens the door and even holds it open for the person. After all, who would stand around a business with equipment in their hands if they didn't belong there? A cybercriminal would!

Emails
The thing that makes email so dangerous is that it is easy to spoof another person with it. The "reply to" setting in an email can say whatever someone wants it to, making it appear to have come from someone else—often a trusted source. As a general rule: the more data points the attacker feeds you, the more likely you will do what they want you to do. Here are two scenarios that can create either a PHI breach or a financial breach:

    i.   The doctor is on vacation and sends his assistant a message to urgently fax the records of one of his clients to him. He needs them! The employee knows the doctor is out of town so it appears legitimate. They fax the information, assuming it's the doctor's hotel, and they never call to verify that the request was really made—the doctor's on vacation, after all. If they had called, they would have prevented a data theft.

ii. The CEO is out of town on business and emails the CFO to wire $45,000.00 to him for a business transaction. The CFO does, but the CEO never requested it. Again, no confirmation and an employee just acting on an assumption.

How do attackers know when people are out of town? They have one simple way to determine it: the email auto reply. *I'm out of town until the 10th and will be returning messages at that time.* From there, it's a matter of simple investigation. Who is the doctor's assistant? Who is the CFO? It only takes a few telephone calls and checking the company's website or social media. Add in that most business emails use individual's names and it becomes even easier to facilitate the rouse.

Hybrid attacks

This is usually a more sophisticated attack because there is usually a specific target in mind. Remote emails and phone calls set the stage and increase the data points for building the trust relationship. From there, physical access based on impersonation becomes more plausible because the attackers have primed the pump. They have badges and credentials around their neck, logoed shirts with a known business's name on them—often a vendor—and it appears that they should be there. And then…the chaos begins.

What is the solution to growing wiser about social engineering attacks? It lies in creating effective and proven policies and procedures. Prevention is as important aspect of helping a business thrive.

PREVENTION:
A COMBINATION OF PRACTICES

Every business should have the goal of preventing social engineering from being a threat to their organization. The necessity for this is easy to understand and embrace, because data breaches impact the bottom line, the reputation, and the success of a business. Here are some steps that I recommend all businesses go through and ensure are in place if they want to stay a step ahead of the cybercriminals who are patiently waiting to expose the "weak link" in your prevention.

1. **Create effective policies:**
   Having policies in place that are signed by all employees is the first step in bringing awareness into your business's operation. Many times, employees just didn't know, but not knowing can be costly

to you—the business owner. As part of these policies you'll want a computer usage section, which should be drafted and reviewed with the help of your IT professional. This is their area of expertise.

2. Have procedures in place for all modes of access:

Whether it is proper email usage or the protocol of how a vendor should check into the building, let employees know what the proper procedure is. Do all visitors check in at the front desk? Can non-employees of the business walk around without an employee with them? Depending on the type of business, these are serious considerations to make and procedures should be in place to maximize the security of your business data. Some smart procedures include:

- Not allowing employees to charge their phones or other electronic equipment through business USB ports, because a cellphone can be an entry point into your system for an attack.

- Never provide more information to callers than what can be found on your company's website, social media pages, and other public access.

3. Employee education and training:

Employees should know and understand what all the threats are to the data that a business has. You cannot have too many eyes and ears alert to possible problems that may be making their way into your organization. This is best done by training employees periodically on the procedures of what not to do, but also what happens if they made an error in judgment. I recommend that you don't make errors punitive, as that may dissuade an employee from disclosing what happened, which will create a bigger challenge to clean up. Plus, it takes away an opportunity to educate.

4. Testing through participation:

It is easy to assume that policies, procedures, and training are enough to create a safer environment for the data your business has; however, this really is not the case. You can gauge how effective these things are by testing through participation. Practical application will always be the best way to see firsthand what employees understand and embrace, compared to what they either knowingly or unknowingly disregard. This step is only to be done after policies, procedures, and training is complete, and

I recommend that every business try this test. It reveals a lot and better yet—it doesn't cost anything.

- Find someone to play your attacker. They'll have to be technical minded and comfortable with this role.

- Have them call your office and say something like: "Hi, I'm Sam from A1 IT Support. We specialize in more affordable IT. If you don't mind my asking, who do you use for technical support?" If the employee doesn't answer—the test stops here. However, if they do, it continues on. Let's say they share the IT companies name and their go-to tech. "We use James from ABC Tech."

- Your attacker will create a fake badge with your IT company's information on it and they'll get to know as much information about the business as possible—either with your help or your actual IT companies assistance for this test.

- A week later, your attacker should go to your work location and say something like, "Hi, I'm Tom from ABC Tech. James sent me over to check out a few errors that showed up on your system. Mind if I go back?" If the employee says no, again, your test is done. However, if they let Tom pass, then it is implied that they have permission to be there.

- Now your attack plant, known by Tom, should ask if he can have passwords for obtaining information and if anyone gives it to them, they should print out everything they can and then leave, forwarding it to you the business owner, along with a report. Ideally—this never went past that initial telephone call.

When this test is complete, it is an excellent source of information for your business to use to further enhance any areas of weakness that were uncovered. This should be done at various times that are unknown to employees, and whenever a new employee is hired, make it a part of orientation.

YOUR BEST PLAN IS TO PLAN FOR THE WORST CASE

It only takes a few well-publicized attacks to make us all more frightened of situations where business data is compromised, as well as our clients' personal data. When this happens, precious resources are wasted trying to correct something that could be avoidable and it becomes a disaster.

Knowing that employees are the last line of defense against attacks is definitely a call to action. *At SkyPort IT, Inc., my staff and I have always worked closely with businesses to develop sound IT relationships where we know the employees and they know us.* We understand each other's goals and purpose. This happens because of on-going team effort of continual improvement. While the security industry cannot guarantee you will not experience a data breach, there are ways to mitigate risk. We can deter attackers from focusing on your business through employee training and layers of protection.

About Daniel

Dan Marcellus knows that technology gone wrong can be downright ugly. As the 17-year president of **SkyPort IT**, Dan helps companies solve technology problems and create systems that complement their goals and infrastructure. Dan and his team at SkyPort IT provide onsite solutions to a wide-range of clients. They've worked on projects like installing new servers for a mid-size marketing firm, helping with equipment for a local robotics high-school team, managing the email services for a manufacturing company, or full-service IT implementation for an ambulatory surgery center. Proper Systems Design and Security are of utmost importance to Dan in proving the best solution – whether it is a fully cloud, on premise or hybrid solution.

Dan's lifelong love for building and programming led to a BS in Information Systems from the Rochester Institute of Technology in 1988. He founded SkyPort IT in 1998 out of a desire to bring old-fashioned values into the IT business: honesty, transparency, fairness, and value. Dan's 30 years of problem-solving experience in the IT field has led to a thorough understanding of the underlying systems that many take for granted. Experience has also taught him that the best way to satisfy his clients is to teach them how to implement better technologies in their companies. He provides seminars, webinars, and original educational materials on his website, and works with each client to ensure that not only is the technology running smoothly, but there are company-wide training and procedures in place to ensure the system continues to function.

SkyPort IT is recognized as one of the best technology companies in upstate NY. Dan's team shines because of his employee-oriented management style, in which he empowers each of his team members to make decisions as if they were the company owner. The SkyPort IT tech geniuses are renowned for their ability to seamlessly take stock of the client's goals, reveal the underlying problems, and create a customized, innovative strategy. This level of personal service has not only led to SkyPort IT's growth, but also an award from a local Chamber, regional press coverage, and hundreds of happy customers.

When Dan isn't solving problems for his clients, he's focused on his family. He enjoys volunteering, playing racquetball, and serving as a Councilperson in his beloved hometown of Lima, NY.

CHAPTER 21

MODERN HEALTHCARE AND DATA SECURITY: PROTECTING ELECTRONIC RECORDS AND SERVING PATIENTS' NEEDS

BY LISA MITCHELL AND JOHN MOTAZEDI

The inception of modern healthcare came in the '60s, and along with it, the requirements for information management grew significantly.

Once upon a time, there was a healthcare system that thrived on paper charts, microfiche, film, and even voice transcription to keep up with the increasing demands of the industry. However, it was not meant to last. Suddenly, new challenges arose, and the management of data became overwhelming to the old systems. The solution was to move to a newer technology; one we know as Electronic Medical Records (EMR).

EMR was meant to bring relief to a bogged down system and create a better one that addressed how:

- Data was stored
- The way we tracked and transmitted data
- And, most importantly, how we protected data

The new era of medicine was sprouting as better systems and processes were put in place. The government also became involved, creating regulations and standardization policies to manage the abundance of healthcare information. The solution was the passage of Health Insurance Portability & Accountability Act (HIPAA) in 1996 which redefined everything. As an extension of HIPAA, the HITECH Act in 2009 helped encourage the use of EMR stemming from the Electronic Health Information Incentive Program, better known as "Meaningful Use". This government program accelerated the adoption of EMR by providing financial subsidies to doctors and hospitals for installing EMR systems and Medicare/Medicaid penalties if paper systems were not replaced with EMR systems. This also furthered the concern for cybersecurity, as a need for it was quickly recognized due to the number of entities that had access to sensitive personal information.

IMPACT ON THE HEALTHCARE INDUSTRY

Medical records are now defined as all documents associated with an individual patient.

Records used to be sent, as necessary, to various healthcare providers prior to HIPAA, via the means that were the "standard of practice" in medical facilities. Consider this:

> Joe is a retired, elderly patient who lives in rural North Carolina. He's in shape and walks a couple of miles a day. He doesn't smoke and isn't overweight. However, Joe has been suffering from migraine headaches and nausea. He has visited his family physician of thirty years multiple times about this issue. His physician is puzzled and recommends a specialist take a look. The traditional healthcare model required a transfer of medical records to the specialist from the physician, resulting in the medical records department spending hours copying all Joe's health information—just for the chance to talk with a specialist. Clearly, a change in the system was needed.

In our new era of medicine, Joe's medical records can be transferred in full, instantaneously. The lag time between physician and specialist is practically non-existent. And better yet, as the specialist meets with Joe, they can see past information, as well as update current information as it adjusts. Joe's medical record is no longer a thick folder filled with paper—it is available electronically and within a matter of minutes.

Today's definition of a medical record is an electronic entity constituting a piece of evidence about the past, which means that a name, social security, birthdate, address, and any individual pieces of data constitute a medical record. Joe no longer has *one record;* he now has *thousands of records.*

With the passage of the HIPAA, HITECH and the Omnibus Final Rule, requirements changed for how confidential information was stored and significantly increased the need for security due to the ease of electronic access to that information.

Now, the entities governed by HIPAA regulations are required to have a comprehensive understanding of compliance with administrative, physical, and technical safeguards in place. These entities include:

1. Health plans, healthcare clearing houses, and healthcare organizations that electronically transmit health information.

2. Business associates that perform services where they have access to Personally Identifiable Information (PII) such as support vendors and medical billing companies.

These safeguards include HIPAA Policies and Procedures, employee workforce training, an ongoing Security Awareness Plan, a Business Continuity Plan for accessing Electronic Personal Health Information (ePHI) in the event of an emergency, a Risk Analysis and Security Management Process, Security Incident Procedures including a Breach Response Plan, a Sanction Policy, device/media disposal, and audit controls, plus other requirements. This is intensive, yet necessary. Prior to these HIPAA regulations, a Physician could guarantee the data was protected because it involved a physical record. Even though the patient folder grew over time, it always remained in the office either on the Physician's desk or the filing cabinet. Things typically didn't disappear – especially without being noticed. Physicians and practice managers would typically say, *"We know where the files are, either on my desk or in the filing cabinet, and we know who has access to them at all times."*

HIPAA regulations give clear guidance as to what steps healthcare organizations must follow to ensure that there is no unauthorized access to medical records. Not having the proper data security protocols in place can eventually result in exorbitant fines and penalties.

DATA PROTECTION FOR HEALTHCARE:
WHAT YOU MUST KNOW

Given the advent of the Internet and the free flow of healthcare information, data security and cybersecurity play key roles in several important topics related to healthcare:

(a). **mHealth (mobile health):** Through the use of mHealth, numerous electronic capabilities and provisions are available, which make medical records more accessible using sources such as the cloud, mobile apps, and Bring Your Own Device (BYOD). mHealth is designed to grant ease of access to medical information and to provide a more responsive patient/care relationship and increase patient satisfaction. This may happen through text, remote access to patient information, and the ability to look up medical records on mobile devices—taking security implications into consideration for public Wi-Fi, Internet, or cell services. For physicians, Joe's medical records are more easily accessed through the use of encrypted-enabled communication for texting, remote access, telemedicine, virtual private networks, and specific encrypted applications for communicating with their EMR programs. This guarantees secure, efficient, and timely communication, but additional security controls like encryption solutions for mobile devices and malware for medical devices must be put in place to decrease the vulnerability of Joe's patient information getting lost or compromised.

(b). **Telemedicine:** Healthcare services for patients in rural areas similar to where Joe lives often rely on telemedicine. Telemedicine allows a distributed Physician Assistants and/or Nurse Practitioners network to see, diagnose, and assist Joe's health care information remotely while using live physician interaction with video, audio, and medical devices in real-time to diagnose Joe's symptoms. Telemedicine brings full diagnostic capability to the healthcare provider while maintaining security and patient confidentiality on both sides. With Telemedicine, important security controls need to be in place including proper access controls—based on the "principle of least privilege" and multifactor authentication solutions.

(c). **Big Data and Healthcare Information Exchange (HIE):** Despite Joe's symptoms seeming simple, he was still referred to a specialist to pinpoint the cause. This specialist is part of an HIE consortium where information is being communicated and shared for the purpose of providing a quicker diagnosis with a more accurate outcome in a more affordable manner, which ultimately leads to better care for Joe. Through the HIE, the specialist can use Big Data Analytics to analyze, track, and diagnose difficult symptoms and diseases. For Joe, this means the specialist can rely on information from similar cases to help determine his needs and define an effective treatment plan. <u>Increasing federal regulations mandate the sharing of information between healthcare organizations, which means that healthcare organizations must have added security controls in place – like Unique User Identification – to ensure the privacy, security, and protection of ePHI and PII.</u>

(d). **Cybersecurity against malicious attacks:** Unfortunately, Joe's medical records and personal data are worth money on the black market to identity thieves and hackers. His stolen records can be sold to the highest bidder, who will use the information to open lines of credit, make purchases, or steal Joe's assets without his immediate knowledge. To ensure that there is no open access to this private information, healthcare practices are expected to have security controls in place such as a proper firewall, Intrusion Detection Systems (IDS), and programs with auditing and logging to protect against unauthorized access to Joe's confidential information. These controls will define the individual(s) who are authorized to view particular information and prevent others who are not directly involved with Joe's care from having access to view his medical records. Auditing opens a window to track and recognize unauthorized data access or security breaches. It is especially essential to have logging controls in place for other areas where information can also be accessed outside of a medical facility such as a secure patient portal.

<u>Cyber Insurance is added protection that many healthcare organizations purchase today. It is designed to protect them against data breaches; however, it is only effective if the organization has a Compliance Plan in place and is following regulations.</u> Otherwise, full claim payouts may not be guaranteed.

(e). Access controls, including a Secure Patient Portal: With Joe living in a rural location, it is convenient for his physician and Joe to share information and correspond through a patient portal system. <u>A patient portal is an interface that allows patients to securely view and gain access to their own information or grant access to an authorized family representative.</u> With Meaningful Use requirements, healthcare organizations are required to integrate a patient portal effectively and safely into their practice operations. Since Joe's information on the portal is publicly accessible by anyone on the Internet, a proper firewall, a secure Operating System (OS), and security patches/updates must be in place to prevent data mining and the compromise of Joe's information by unauthorized individuals. Please note— patching and maintaining an OS and firewall is an ongoing process. Proper firewall management provides better protection against intruders, malware, and viruses. Plus, access will be logged that clearly defines the window of exposure, should an unauthorized person gain access to confidential information.

(f). The cost of a security breach: Lapses in security measures can quickly cause irreparable damage with financial penalties, reputational harm, and loss of business. If Joe's ePHI is breached, healthcare organizations are potentially looking at the following consequences:

- Costs for excessive legal fees
- Public relation expenses
- The loss of business
- Expenses for assessments and ongoing investigations to protect the organization
- Setting up and maintaining a mandatory Compliance Program
- Credit monitoring services for consumers like Joe

In addition to those costly business exposures, healthcare organizations may also have to devote time and resources to develop their Written Information Security Policies (WISP), which define a set of procedures and policies that ensure the confidentiality, integrity, and availability of the organization's information against malicious activity. When developing cybersecurity and HIPAA compliance programs, organizations must pay particular attention to the following:

Other federal and state regulations in addition to HIPAA requirements in regards to ePHI and PII.

Data breach laws for mental health and substance abuse, because there are specific guidelines established to protect against how and when ePHI and PII are shared in these situations, whether with or without the patient's consent.

(g). Security Incident Management: Joe's information has been compromised. Now what? Is your staff confident that they know how to handle this breach? An organization never wants to wait until they are in a breach to know what needs to be done. Here are twelve steps for handling a security breach:

1. Contact a HIPAA Consultant and/or a HIPAA Attorney.

2. If over 500 records have been compromised, provide notice to prominent media serving your state or jurisdiction by the deadline.

3. Notify the patients affected by individual notice by the deadline. Apologize for the breach and indicate that their medical information <u>may</u> have affected.

4. Implement your Employee Sanction Policy and document the breach in the employee's personnel file.

5. Complete a Security Incidence Report Form with the supporting documents.

6. Review a Data Breach Notification and Mitigation Checklist.

7. Determine if more or less than 10 contacts were invalid and if a Substitute Notice is required, either by posting the notice on the homepage of your website for 90 days or by providing the notice in major print or broadcast media where the affected individuals likely reside.

8. If required, provide a toll-free number for 90 days where individuals can learn if their information was involved in the breach.

9. File a Data Breach Report with the Office for Civil Rights (OCR).

10. Review the breach, mitigate the circumstances, provide education around the breach event, and re-train staff.

11. Review federal and state breach laws for additional steps, especially concerning PII.

12. Follow both federal and state mandated laws, including potentially notifying the Attorney General and/or the three major consumer reporting agencies.

YOUR GOAL IS BREACH PREVENTION

No organization wants to find itself in the position of having to defend a breach.

Preventing a breach is ideal and being proactive is necessary. Take these ideas under serious consideration:

- Hire a Managed Service Partner to monitor and manage your IT environment.
- Have regular consultations with a HIPAA knowledgeable attorney.
- Hire a HIPAA Consultant.
- Pay a professional to conduct a Comprehensive Risk Assessment.
- Mitigate the results of a Security Vulnerability Assessment.
- Implement and enforce HIPAA Security Policies/Procedures.
- Get Cyber Insurance and implement a full HIPAA Compliance Program.
- Invest in industry-standard encryption solutions.
- Have an Organizational Policy for addressing and monitoring mobile devices.
- Use Active Directory and Group Policies to enforce User Rights and Security Controls.
- Ensure that medical devices have up-to-date security patches and are malware-free.
- Implement and maintain strong firewalls with subscription services and an Intrusion Detection System (IDS).
- Cultivate a strong, ongoing Security and HIPAA Awareness Program.

Cybersecurity enforcement plays an instrumental role in ensuring that an organization's most valuable asset, the critical data, is safe. Protecting ePHI and PII is critical with the passing of the Omnibus Final Rule. The stakes have been raised substantially and there are serious consequences when HIPAA regulations are violated, including fines that few organizations can afford.

HEALTHCARE ORGANIZATIONS MUST TAKE AND MAINTAIN CONTROL

Taking for granted that your plans from last year are effective this year can be a costly mistake.

Much like HIPAA, every organization also needs to continually assess their level of risk for security, evaluate the maturity of the current Security Program, mitigate any ongoing security concerns, implement their WISP, and develop an ongoing level of security awareness. From there, you need to continue to test and reassess the current security measures in place.

It's true, even the most cautious healthcare organizations can experience data breaches. It's as simple as having one disgruntled employee, a dissatisfied patient, a conscientious employee who takes work home to finish, an intern working for the summer, old users' access rights not properly terminated, copiers not being wiped before being disposed, or an office break-in where equipment is stolen. Organizations should always know and double check where their sensitive data resides, whether it's on local workstations, mobile devices, thumb drives, scanners, copiers, etc. Most importantly, protecting your data is the best way to avoid a breach. In addition, training staff to fully understand HIPAA and the value of electronic information is necessary—but it pays off! Ultimately, a healthcare organization's protection is dependent upon having an informed and trained staff. Organizations that take these proper precautions have a much better chance of avoiding excessive fines, surviving a breach of confidential information, and ensuring that Joe's medical history remains safe and protected.

About Lisa

Lisa Mitchell is a graduate of The University of North Carolina at Chapel Hill and she is the CEO and co-founder of Progressive Computer Systems, Inc. Lisa and her business partner, Mark Michal, started Progressive Computer Systems, Inc. in 1987 with the sincere desire to cultivate long-standing partnerships with both their clients and their staff based on building trust and operating with integrity at all times. Many of their clients have been with Progressive for over 15-20 years and the average employee retention rate is over 12 years – which confirms one of Lisa's core beliefs that your team is your most valuable asset.

Lisa is passionate about helping her clients become proactive in managing their businesses and providing Strategic Planning to ensure their future successes. By serving primarily as a high-level Virtual CIO (Chief Information Officer) for her clients on multiple fronts, Lisa is able to help her clients not only make good decisions around Information Technology (IT), but also help them focus on the broader components of what makes a business succeed in general. While the core services that Lisa's company offers include IT Managed Services, Cybersecurity, and HIPAA Compliance, what makes Lisa's knowledge valuable to her clients is the business acumen learned in the almost 30 years of running her own IT business. Lisa also currently serves as the facilitator for a national Healthcare Special Interest Group and she is involved with other International IT Peer Groups and local, non-profit organizations.

Lisa and her team have been fortunate to have received national and local recognition including being rated as one of the "Top 500 Technological Companies in the Country," one of the "Top 100 Technology Companies in the Triangle," twice recognized as the Top 50 Fastest Growing Companies, Top "40 Under 40" Businesses, and one of the "Top 25 Women in Business." Progressive's clientele ranges from small, medium, to enterprise-level clients in all professional industries with a heavy concentration in the medical, pharmaceutical, dental, legal, and healthcare vertical markets.

You can connect with Lisa and Progressive Computer Systems, Inc. at:
Lisa@pc-net.com
www.pc-net.com
www.facebook.com/progressivecomputersystemsinc
www.twitter.com/progressive_cs

About John

John Motazedi is the CEO and founder of SNC Squared. He started the company in 1998 as an individual, and SNC Squared has grown to a staff of fifteen.

He is a leading expert in disaster recovery and has been recognized by Fox Business, NPR, PBS, CRN, FedEx and the Missouri Health Information Technology Assistance Center for his expertise and swift response during the May 2011 tornado that devastated Joplin, MO. He has been a guest speaker at multiple regional and national disaster recovery events. He is also a leading force in helping businesses in the Joplin Metro area develop their own disaster recovery plans. In fact, he has best-selling book on the topic of disaster recovery titled, *The Tech Multiplier.* It is available for purchase on Amazon.com.

John was named a recipient of 14 for 14 from the Joplin Metro Magazine that chronicled fourteen individuals that will helped define the Four-State area in 2014. John was also the 2012 winner of the Technology Marketing Toolkit, which awards the best IT company in the US, UK, Canada and Australia. SNC Squared was voted the number one company by Heartland Technology Group for efficiency, growth and best delivery of solutions to the end users. They were also awarded the highly coveted *Hands that Give, No One Left Behind* Award for reaching out and assisting during the time of need in a disaster.

SNC Squared was named Small Business of the Year 2013 by the Joplin Area Chamber of Commerce and was one of the top finalists in 2012. They have been accepted in the 2012 and 2013 Edition of Inc. 500|5000 Magazine for being one of the fastest growing companies in the nation.

Additionally, SNC Squared was recently ranked at number 33 for the annual list of the top 500 fastest-growing Ingram Micro U.S. channel partners.

You can connect with John and SNC Squared at:
John@sncsquared.com
www.sncsquared.com
www.simpleithealthcare.com\
https://www.facebook.com/sncsquared2

CHAPTER 22

A MILLION DOLLAR QUESTION: IS YOUR BUSINESS IN COMPLIANCE?

BY MARK BOWLES

When it comes to government regulation for data security, credit card processors, publicly traded companies and the healthcare industry face severe penalties for non-compliance. The Health Insurance Portability and Accountability Act (HIPAA) is perhaps the most far-reaching – affecting the over 230,000 medical practices in the United States.

To consumers, HIPAA (Health Insurance Portability and Accountability Act) is most likely nothing more than another acronym for regulation that they know protects them in some way. But do most consumers understand it? Typically, no. In fact, most people have no reason to think about HIPAA until they are alerted that their very private information has been compromised. **Then, they have a problem**!

What we really need to evaluate is: <u>Do businesses that work in the healthcare industry, in some way, fully understand HIPAA and its security rules</u>? Most of what is required to stay compliant with HIPAA is the full responsibility of the organization that collects the information to perform a certain service. If you are a business owner, are you compliant? And, if you are a consumer, is your medical office and its vendors compliant? *These are the questions that need answers.*

WHAT IS HIPAA'S INTENT?

Consumers deserve and demand privacy and security rules with their personal information.

Upon its inception in 1996, HIPAA became the act that **protects** *the privacy and security of individuals' health records*. Its intent is not to block any individual from having their information shared as they see fit. Rather, it is to prevent people from gaining access to it that will not use the highly-confidential information for its intended purposes—improving and maintaining your health. Did you know:

Medical records are one of the most valuable items that a hacker can get a hold of—being worth up to several hundred dollars apiece on the black market.

The wrong hands gaining access to your medical records gives a criminal access to your entire life…and they will use it to their advantage, not caring how it destroys the "real you." Medical records include:

- Full name
- Birth date
- Social Security Number
- Phone numbers
- Employment information
- Current address
- Medical history
- Personal and emergency contact information
- Possibly credit card information
- Prescription history
- Your marital status, and even the number of children you have

Ask yourself. . . <u>Would you ever hand this information over to a stranger</u>? Of course not! **Our healthcare providers are people we trust and rely on to increase the quality of our lives,** or manage the conditions we face health-wise. We are willing to provide what they need in order to receive the best care possible. That is the heart of this act. *HIPAA is your protection that will lessen the odds of your valuable information being exposed to the wrong person.* It's your defender against crime, but it's the technology of savvy professionals who are the super hero, because they have the ability to put protections in place that prevent a thief, even in the virtual world, from becoming you.

The first HIPAA regulations came out when electronic record sharing was starting to boom and be considered for all businesses, not just medical practices. In 2003, when I founded Ranger Solutions, I established many business relationships with various types of medical practices and networks. I knew right away that understanding HIPAA was something that was necessary for my clients' success and compliance. I read the act cover to cover, making sure that I was doing everything possible from my end to ensure I was guiding clients in the right direction. Because if I didn't, I was increasing their chances of being vulnerable to civil penalties, criminal penalties, and of course, hackers.

My clients that own medical practices appreciate that they can rely on me to help protect them. I've proven that I can do that effectively, and when I get the chance to meet potential new clients or just a curious person, they often ask, "What's the number one thing a business can do to protect records?" The answer: *Have the right technology measures in place!*

HOW DO CONSUMERS KNOW WHICH MEDICAL-RELATED BUSINESSES ARE TRULY HIPAA COMPLIANT?

Every doctor will say they are compliant, because they usually believe they are.

The owners of businesses that are related to the medical field in some way, whether it is a doctor's clinic, a transcription service, insurance processing, etc., are the ones who are responsible for ensuring they are compliant with the specific guidelines. There isn't a lot of room for leeway in interpretation, and that is to everyone's benefit.

If a business wants to ensure they are as HIPAA compliant as possible, an assessment from a qualified IT professional is an excellent starting point. Our firm, Ranger Solutions, does this type of assessment and it has proven to be one of the best assurances that we can give our clients, who are committed to protecting their clients' privacy, as well as their business' integrity.

Fortunately, becoming HIPAA compliant is not a process that requires a medical office to carry the expense all on their own, either. This is a great incentive! *The government offers these businesses "meaningful use," which are funds allocated to medical practices for assisting in upgrades to computer equipment and software, so that they are in HIPAA compliance.* <u>If a business ever fails their review at any point, they are no longer eligible for those funds</u>. I'm sure you can understand why they would prefer to have those funds coming in, instead of paying out for these upgrades (which will have to take place) themselves.

There are also two interesting points about HIPAA compliance that consumers may have noticed over the years, as the culture of medical practices has shifted.

1. **Requests to turn your smart phone off**.
   You may have thought that medical offices post this because they want to caution patients to not be a nuisance to others by talking on the phone and disturbing people around them. That isn't the true reason for this request, however. This is posted for patient privacy, which is a key component in HIPAA. For example: If a patient were to photograph another patient and it's uploaded to social media, the photographed person could file a complaint that their privacy was violated, which would lead to an investigation of the practice by Health and Human Services. It would be the medical facility that was investigated, not the consumer who took the picture. <u>Privacy is taken very seriously</u>!

2. **The "Wait Here Until the Patient Ahead of You is Done" sign**.
   Long ago, you'd show up at the doctor's office and sign a piece of white paper that was up at the receptionist desk to let them know you were there. This is no longer allowed for privacy purposes. Instead, the sign indicates someone checking in for an appointment to wait behind a certain point while the person ahead of them finishes. It's to protect their privacy. Since they may be disclosing personal information—such as insurance, phone number, a medical condition—they have a right to be safe from prying ears.

Would you ever have thought that so many people played a role in HIPAA compliance? Perhaps you were surprised to find that you had a role in it, as well, from a consumer level, although it's the businesses responsibility. Everyone, aside from hackers and criminals, has the same goal: *to receive necessary services without the risk of personal compromise.*

WHO IS RESPONSIBLE FOR WHAT WITH HIPAA?

There is no "passing the buck" with HIPAA compliance.

The owner of a business is the one who is responsible for ensuring that HIPAA guidelines are being followed to the utmost potential. However, they do rely on their support staff for much of this. Here are three bits of wisdom to help gain a better grasp of what can be done to make sure that everyone is paying attention to, and is acutely aware of, compliance guidelines.

1. **Saying, "I don't know," means penalties**.
   Violations can be either a civil penalty or a criminal penalty. Civil penalties have four tiers, which are:

   - Tier 1: Covered entity or individual did not know the act was a HIPAA violation. The fines are per occurrence, ranging from $100 to $50,000 per identical violation during a calendar year.

   - Tier 2: The HIPAA violation had a reasonable cause and was not due to willful neglect. The fines range from $1,000 to $50,000 for each identical violation during a calendar year.

   - Tier 3: The HIPAA violation was due to willful neglect, but the violation was corrected within the required time period. For this, fines range from $10,000 to $50,000 for each identical violation during a calendar year.

   - Tier 4: The HIPAA violation was due to willful neglect and was not corrected. Fines for this are $50,000 or more for each similar violation.

   Although the fines for these violations are capped at $1.5 million dollars for same-type violations, you can see that it is a costly endeavor to say, "I didn't know." When it comes to criminal penalties, the price of violating HIPAA could be your freedom.

   - Level 1: If you are non-compliant, either unknowingly or with reasonable cause, you could face up to one year in jail.

   - Level 2: If you willfully act under false pretenses, you could face up to five years in jail.

   - Level 3: If you use private information for personal gain or malicious reasons, you could face up to ten years in jail.

For the record, I've heard of only one case in which a medical practice was blatantly malicious, selling patient information for profit, so Level 3 is the exception to the rule. Still, mistakes can be made when assumptions are made.

2. **Have ongoing monitoring in place to ensure that guidelines are being followed.**
These measures include business firewalls, monitoring that is specifically geared to identify potential hacks and break-ins, and making sure files are backed up so you have at least seven years of all your patients' records. If a consumer requests this information, you are obligated under HIPAA to provide the information to them within 30 days.

3. **Require online training for support staff, as well as the business owners.**
There are cost efficient—considerably less than $1.5 million dollars—online training courses to ensure that everyone that works for you is HIPAA trained. They will receive a certificate of completion, which will give you assurances that they understand the importance of HIPAA to their jobs, as well as to the business' livelihood. Also, medical businesses should ensure that their vendors are HIPAA trained, as well. You cannot have too much precaution or education with this act.

WHAT HAPPENS DURING A HIPAA EVALUATION?

The entire purpose of an IT HIPAA evaluation is to ensure that you are operating from the most compliant position possible.

Ranger Solutions clients' look at us as the trusted partner. This is because we have helped them prepare for HIPAA and to better understand everything that is required, ensuring that procedures are being followed and maximized through the types of technology that are in place. We look at these five core areas to make this determination:

1. **Ensure the right documentation is in place.**
This has to do with a business's IT plans, and even their emergency back-up plan, should they have a disastrous event, such as a server crashing. Regardless of the chaos that may exist during one of these events in the physical environment of a medical practice, the

electronic environment must be protected because of the valuable data it contains.

2. Have smart user policies.
Make it a requirement to change passwords frequently, and also know which staff members have access to what data. You need to have a log that shows what employee was in what online area and when. Doing these things demonstrates that you're taking a professional and educated approach to HIPAA compliance, and are focused on using the best experts to help reduce your risk of non-compliance.

3. You cannot have out-of-date software and hardware.
This is best explained by example. Today, any medical practice that runs Window XP or Windows 2003 Server is in violation of HIPAA because they are no longer being patched and maintained. Security updates must be able to be performed.

4. Perform updates and back-ups on a regular basis.
In-house IT is not the most cost-efficient option for many businesses, but hiring a business like Ranger Solutions is. This is our fulltime occupation, and basically, what we live and breathe. Doesn't it make sense to have multiple eyes and people with specific IT experience looking out for you and your business? We are constantly learning and need to stay one step ahead of cybercriminals and technology changes. That is why our clients call us partners and friends, not just "the IT guy."

5. Make sure you have accessible log audits.
Many businesses do not need to keep their log audits, but for medical businesses, it's required for compliance. They must be archived and be easy to pull up whenever necessary. For us IT-minded people, it's our way to watch over what is happening within our clients' businesses and to make sure that we are achieving the desired outcomes. Plus, it is gives us a greater opportunity to look for patterns and draw information from other cases that may help our clients.

Very few things are ever "one and done," especially HIPAA and IT. It's ongoing updates, finessing, and understanding the changes that are taking place in laws, acts, and consumer patterns on a regular basis. Even HIPAA guidelines change—they have changed twice since their inception, with the latest adjustments taking place in just 2013. *These are things we stay 'on top of' so we can be proactive with our client—which benefits them, consumers, and the integrity of the medical professional in general.*

WHAT'S NEXT?

Sometimes the best defense is a good offense.

If you are a consumer, <u>don't be afraid to take the time to make sure that your medical providers are following HIPAA guidelines.</u> Ask questions! It's okay to do that. They've likely offered you a privacy notice that has a recap of the procedures they have in place. But have you read it? Take the time to read it, because it's written in consumer terms.

For businesses, you have worked so hard to build relationships where people either trust you with their health, or health providers trust you with their clients' data. **Being HIPAA trained and compliant isn't just an accolade, it's a necessity**. Assuming that everything is in place is a philosophy that leads you to finding out that you did too little, too late.

In the end, we all have certain expectations about our information that we share and how it's managed. **HIPAA offers some peace of mind,** <u>because it helps ensure that those who are in the business of helping others continue to thrive, while eliminating risks from those who may be focused on criminal mischief</u>.

About Mark

Mark Bowles helps his clients secure their business in the online world. With over 30 years' experience in operations management, Mark brings a wealth of expertise in harnessing the power of technology to achieve business goals. He served in executive roles for a premier provider of outsourced managed computer services. He created and led a technology team that designed and implemented over 400 custom infrastructure solutions that accounted for a majority of the provider's revenues. In 2000, Mark's team won the Microsoft Global Service Provider Award for Hosting.

Mark founded **Ranger Business Solutions** in 2003 to address the business technology needs of mid-size companies. In his previous roles, Mark led multiple teams as a "Big Eight" business consultant, spearheading technology implementation projects for Fortune 500 companies in the telecommunications, manufacturing and consumer products industries. Today, Mark's firm provides ongoing business technology support to over 100 businesses across the United States.

Mark is a graduate of The University of Texas at Arlington. He is the CEO of Ranger Business Solutions, a Managed Services company, specializing in business technology solutions.

You can connect with Mark at: Mark.Bowles@RangerSolutions.com

CHAPTER 23

WHY A UNIFIED CLOUD SOLUTION MAY BE YOUR BEST OPTION TO COMBAT CYBER CRIMINALS

BY STEPHEN KUSKA

Imagine working for "Company A" and coming into work on Monday morning, the week of the big event. Two thousand people are attending a semi-annual conference in Orlando at the Peabody Convention Center. With three days to go before the first session, you log into your computer to start printing the materials for all the attendees. You have gone through this process twice a year for the last four years, but today all the files on your computer appear to be missing. You attempt to look for files on the server, and they too are gone. You check a third computer with the same results. The panic has finally set in… where do I go from here?

The exponential proliferation of Internet-connected devices has created an increased need for security solutions to protect those devices. Normal development time to engineer solutions to security issues can take months. As a result, the newest technologies are the same technologies that are most vulnerable to cyber security threats. Consumers and business owners are now faced with the problem as well as the responsibility to find ways to protect their identity and data.

When you start to consider "The Internet of Things" the situation becomes even more complex. Almost any device that is connected to the Internet can serve as a potential access point to your data. Day-to-day devices, such as cars, refrigerators, smoke detectors, video cameras, televisions, security systems have become entry points to all networks. Chinese hackers have already successfully hit the US Chamber of Commerce through a thermostat installed as part of a new air conditioning system.

On the surface the problem seems monumental. There is an ever-increasing number of devices that require a growing number of software applications to protect them. . . at a cost that is also soaring. This trend does not seem to be changing anytime in the near future, and unfortunately, there does not seem to be a silver bullet on the horizon that will fix all our problems. It is time to focus on how we can mitigate our exposure without giving up all our new toys.

It is important to understand that the real problem is that cyber criminals want our data in order to exploit our identities, and our access to financial accounts or even the accounts of our clients. Much of our personal data resides on the devices such as our computers, laptops, tablets, and phones. In order to mitigate the loss of this data, we need to be able to control the number of places that data is stored. Take into account that every employee within your company has access to data in multiple ways, the problem becomes even more self-evident. Each employee has their own access through multiple devices and pathways, some of which are not even onsite. Each of these entry ways can be an access point for a breach into your system. The question becomes: How can I protect all these places?

The task becomes much easier to manage if there is only one place to protect, backup, or update software. The decision to centralize and remove the data from our devices seems to be the obvious choice. Once the decision to centralize has been made, you need to start looking at how to implement this solution.

Since the ability to communicate to the end user is the main requirement, it seems to make sense to at least consider the Internet (Cloud) as the provider of this resource. Internet technologies are being implemented on a worldwide basis and currently provide the majority of all communication in use today. Everywhere we go today we hear about the cloud. Reliability issues seemed to have disappeared, and costs have come down to the point that it is now affordable to have multiple vendors providing the same service to meet redundancy requirements.

So let's start by taking a look at a small company of about 25 users. A company of this size normally has 2 to 3 servers, with 25 workstations. Factor in that each employee will also use a cell phone and tablet to process email and other company data. We now have over 75 devices which need to be protected. Since the data that needs to be protected is located on the network, the in-house IT staff must install the technical solutions for all the different devices that are attached to the network. They have to look at all of the firewalls, workstations, routers, PC's, laptops, and telephones.

This year's Internet Threats Report from Symantec reported there were 317 million new types of malware discovered. The bad news in this report is that the malware is getting much more sophisticated and in many cases can evade detection. Current antivirus programs do not seem to have a chance to detect these newer sophisticated threats. Once the threats make it to your computer, Smartphone, or tablet, they can introduce some very unwanted behavior. But rest assured, there is a multitude of companies trying to provide solutions to address all these platforms.

Since software vendors are required to provide the federal government with information about software vulnerabilities as they find them, the federal government publishes, on an almost daily basis, a list of vulnerabilities that IT departments should be on the lookout for.

Unfortunately, there are software products available that will allow an external scan of a company's network and provide hackers available ways to penetrate. The vast majority of all system penetrations are through previously defined and documented vulnerabilities to systems that are not properly maintained. It becomes clear that because of the ever-expanding numbers, threat management has become a full-time job, often requiring more than one individual for a particular network environment.

From the cyber security aspect, cloud providers seem to have the upper hand. When done properly, infected devices cannot directly connect to the cloud servers, thus providing a barrier from unauthorized data access. Cloud providers have the capability of verifying the identity of those trying to access the resources not only by user ID, but also by location, or device ID. Cloud providers are currently ensuring that only those resources that have met security standards such as an active and current virus software running on their devices can access data.

They live in a very controlled environment where change is done by exception. With full time staffs to review threat publications, they are positioned to implement fixes across their entire client base as soon as they are identified. When you add their capability to do weekly scans of all their systems and their requirements to undergo periodic security and compliance audits, they currently seem to have the upper hand when it comes to implementing security procedures.

Although threats come in many sizes and forms, in the unlikely event an incident should occur, most cloud providers are in a much better position to respond because of the infrastructure resources they utilize. Infrastructure design used to support these facilities are normally outside the financial resources of small and many mid-sized companies. Recovery tasks that are currently considered mainstream may require days to replace equipment and restore backups. These same tasks can be accomplished in minutes with the infrastructure currently in place in most cloud datacenters.

Recently, a firm of significant size was unfortunate enough to receive a copy of the "Cryptolocker Virus" at one of its remote locations. As with the example above, the virus sat in the client's computer and systematically followed all the links on that computer to encrypt all the files it could find. The net result, similar to the example above, was that all the files were missing. When an attempt to access one of the infected files was made, a message popped up requesting a ransom in order to provide the mechanism to restore the files. Although all the files were backed-up off site, by the time one considers travel, virus removal and restore time, in excess of three days were accumulated before everyone was back in operation. In an adequately-hosted environment, the systems could have been restored in approximately 20 minutes.

Until now, we have mostly talked about the mechanics of the cybercrime problem and solution, but a unified cloud solution has the power to provide so much more. This is where the real cloud computing excitement is happening.

Moving all your applications to a unified cloud solution can make everything IT simple. It not only includes all the cloud computing benefits we talked about above, but it offers so much more.

For your staff, a unified cloud solution provides integrated web-based access to all of their applications, data and email, and gives them integrated file sharing and collaborative tools. It allows them to work just as easily with Web-based applications as it does with all cloud-based applications. And it's right at home on a PC, Mac, tablet or mobile device. From any Internet link in the world, they can have single sign-on access to all of the IT assets presented on the same desktop that they use when they're working in the office.

Employees can post announcements, create shared departmental files, reference an automated, up-to-the-minute company directory, and tailor WebTop to add links to their favorite web sites.

For clients or business partners, you can provide secure, audited email and/or file access to your servers. As for email retention, you can have it archived automatically for up to ten years.

For company administrators, a portal desktop is the world's fastest, easiest way to manage IT assets. Let's say that business is good and you need to add someone new to your staff. In the new environment, there is no need to call IT any longer. Setup can be completed through a set of easily-navigated web panels, and all the infrastructure is complete.

You can easily control permissions or add and delete users, user groups or departments. With little effort you can update software, grant access and oversee security. An administrator can even train someone new (or several people) in just an hour.

As you can imagine there are many companies that are entering this space in the market. There are companies such as Google, Microsoft, Apple, Amazon, and a host of others providing services to select segments of the consumers. One must take care however, to ensure that whatever solution you choose, it meets all security and retention requirements for you or your organization. For these reasons, I would recommend looking toward a fully integrated and tailorable solution as provided by a company such as OS33.

Another part of security that is often not considered until it is too late, is the ability to recreate who, what, where, and when. Unified cloud solutions include the reporting from the beginning and allow for audit trails for all company activity. This can be a real asset should a company need to retrace company activity.

When it comes to IT and data processing, we are in the middle of a paradigm shift. We can no longer spread the data we need to protect to every device we touch and expect it to be protected. Placing that data in the cloud with a provider staffed to focus on data security, with the resources to recover quickly should a failure occur, is the best of all worlds.

If you happen to have an IT wish list, a unified cloud solution will let you check off all of your boxes. What's more, moving to the cloud with the right solution won't take years or cost a fortune. In fact, it's designed to save you money from the very first day. Capital expenses? Goodbye. Buying additional servers? Goodbye. Worrying about issues like backup, security, and on-call support? "Adios! Sayonara! Farewell!"

Let's take one last look at "Company A." At the time of their problem they were a small company with eighteen employees. Shortly after their panic attack, they made the decision to move to the OS33 cloud. Three years later, they have increased their staff five fold, across four continents, many countries, and all are operating on the same set of applications and data at the same time. They have integrated products such as Sales Force and Office 365 with their local applications, have single sign on capabilities for all their total environment, and currently do not have to worry about the type of problems they have experienced in the past.

About Stephen

Stephen Kuska has over 42 years of experience in the computer consulting industry. His early years (15 Years with IBM) gave him the opportunity to be involved with industry-leading technologies from incubation to delivery. Working with some of our nation's largest companies, Stephen was instrumental in implementing these emerging technologies which gave those companies their competitive advantage in the market place.

After leaving IBM in 1987 and starting his own consulting firm, Stephen continued his relationship with IBM. Working in a partner relationship, Stephen's firm was instrumental in completing a number of contracts/projects with many companies that include American Express, Royal Caribbean Cruise Lines, Office Depot and Motorola. After a number of years operating in this environment, Stephen's company was recognized by IBM for their contribution in assisting them achieve their *Malcolm Baldridge Award* (the only National Quality Award.)

As President of **Computer Solutions Unlimited**, Stephen remains engaged in developing future technologies. To date, Stephen has been able to lead his clients through the age of the mainframe computers, through the PC distributed network age, and is now well-positioned to lead them through the next generation of cloud computing. Drawing on his years of experience in the ever-evolving technical environment and his focus on the security and cybercrime issues of the day, he is in a unique position to address current network security situations, and their challenges.

As the only OS33 Certified Partner in Florida, he brings the same leading technologies in use by companies such as Fidelity Financial to the local marketplace. He is well-positioned to speak about the future and address how organizations can best solve some of today's challenges.

CHAPTER 24

LOCKING OUT CYBERCRIMINALS: THE SOLUTION IS LAYERS, NOT WALLS

BY DANIEL FOOTE

Your systems <u>are</u> under attack and <u>need</u> to be protected.

There was a time in your life when you thought your biggest gamble was going into business. It's a huge risk, after all. However, you went for it and it's worked out pretty well, which is great. Now you are facing another risk that could threaten to take away everything you've worked for. <u>And you've worked hard</u>! If someone were to ask you how assured you were that you had the right layers of security in place, would you be confident in your response?

As a business owner, look at the three statements below and see if any sound familiar to you. Have you said them? Have you thought them?

- "We don't have anything that someone else wants."
- "We're too small; nobody's interested in us."
- "We've got nothing to hide."

Believing that there is validity and truth in any of these three statements is a red flag for your business's well-being. These statements are false assumptions and if you believe them to be true, you are likely experiencing systemic attacks on your network daily. Is that really what you want? Of course not! Here's what our 15 years of experience has proven to be true for business owners:

- Perceived "unimportant" data is used by cybercriminals as a means for them to create a financial gain—for them, not you!

- Small businesses are ideal targets, as they "feel insignificant." But, they are not!

- Compliance requirements for business transactions are getting stricter.

- Security isn't about "hiding;" it's about "protecting."

YOU'RE AN ORIGINAL; DON'T BECOME A STATISTIC BY OVERLOOKING PREVENTION.

There is no "one-size-fits-all" IT plan for creating a more secure business network.

Every business is different, and has different needs. Compared to small businesses, banks and financial institutions employ complex methods to ensure the highest levels of security available. Hospitals and large healthcare organizations choose different security techniques from what a small clinic might. Regardless of the size of the entity, the approach comes down to one factor: the ability to allocate resources plus the staffing to monitor the systems necessary to manage all technology that has been implemented.

Larger businesses aren't invincible, but overall, they're better protected than the smaller businesses that may be under the illusion that they are insignificant. Think of a speck of sand in the desert. At first glance, you think that it blends in and could never be singled out. Guess what? Cybercriminals are experts at finding that speck of sand and exploiting it to a high degree. Our understanding of this is a large reason why DanTech Services is quite passionate about educating and informing smaller businesses about data protection and intrusion prevention—no exceptions to the rule!

WHEN TECHNOLOGY AND HUMANS COLLIDE . . .

Data security is a partnership between the employees of a business and the technology that has been put into place.

The relationship between a business's employees and its technology is very important. You can have the best technology in place, but if your employees don't understand their role in protecting its integrity, it's incomplete. Likewise, you can train your employees on what to do and not to do with technology in the workplace, but if you don't have excellent security measures in place, your data is still highly vulnerable. Nothing better emphasizes this point than the following situation, which really happened:

> Phil is an IT specialist and he goes to perform a routine audit for one of his clients. He enters the building and is glad to see that everything is as it should be: he has to sign in at the front desk and cannot gain access to anything confidential without the appropriate management's presence. It's ideal, and he's thinking, this is great, a real success story. He walks through the building and sees that all suggestions are in place and continues to be more impressed. There's only one check left in the physical security layer, which is the perimeter of the building.

> Once outside, Phil walks to a building next door and goes to the roof to assess his client's building from a different angle. This is when things get interesting. He sees a fenced area in back of the building. Inside this 'protected' area, there is a door open that leads inside. There is also a ladder, in the grass, next to the fence. Once the ladder is moved against the fence, anyone who can climb a ladder can now scale the fence. Red flag! Curious as to what he may find, Phil leaves his observation spot and makes his way over to the fence, uses the ladder to go over the fence, and effortlessly walks right into one of the most highly secured areas of his client's business.

> What's the moral of this story? *The most advanced strategies and systems can be overridden when common sense doesn't prevail.* Never overlook the obvious, because data thieves do not.

Data technology relies on employee awareness to help prevent costly data breaches. You must train employees, as well. Employees must understand the risks of cyber security and the importance of the data they have access to. This is true for data already stored on a system or if it is incoming via emails or data entry. Here are two things that humans can control for better data security:

1. **Business owners should hold qualified security awareness training for employees.**

   This type of training is customized to the type of business and it includes learning to recognize phishing (posing as a credible business online to gain information from a consumer or business), and spam (messages sent to large numbers of people on the Internet in hopes of acquiring something). This may seem overwhelming, but here is the simple solution: *If anyone is unsure of an email being safe, ask an IT person first—before clicking on it—and you will greatly reduce your risk of a malicious link worming its way into your system.*

2. **Have an Acceptable Use Policy (AUP) and Internet Access User Policy (IAUP).**

   Since staff carries out the day-to-day operations and helps a business function, it is important to give them information on exactly what is acceptable and not acceptable with technology. Disgruntled employees on a poorly-protected system have a plethora of opportunities to sabotage you before they either make an exit of their own accord or are released from their positions. They could: alter records that are in the system; take data and give it to a competitor; or use the data as a means of gain by selling it for money. It's disheartening to think that these things can happen, but they do. Being realistic about it is best. Create evidence that employees understand what is expected of them. This is done through having a signed AUP and IAUP, which Internet security specialists can help you craft. These pieces are also the perfect companions to layered security. Also, keep in mind that every employee does not need access to all the data you have. If it doesn't pertain to their job function, don't give them access. And be devoted to the practice of changing passwords several times a year.

DO YOU HAVE THE BIG SIX TECHNOLOGY LAYERS WORKING FOR YOU?

When cybercriminals target your business,
force them to take a pass.

There are strategic technology tools that a business can implement, which will provide peace of mind to the owner and deter most cybercriminals, forcing them to seek unprotected targets.

Is there ever a 100% assurance that a business will not be hacked? **No, there isn't; however, there are ways of minimizing the odds to practically non-existent.**

1. **Contain what already exists.**

   Have a perimeter and network security check done and act on the recommended adjustments immediately. This includes ensuring that the system is properly configured, managed and monitored with a Next-Generation Firewall, or Unified Threat Management appliance (UTM). Consumer grade equipment should not be used for business security. With a UTM, you'll be provided features such as a gateway antivirus, intrusion prevention service, application security, virtual private networks and access controls. This is also where managed network switches can be configured to allow only known devices to access resources or to segment access to data, devices, and resources through Access Control Lists (ACL). Does this sound confusing? It won't be if you have the right IT specialist by your side. The best in this field have an ability to make this "must know" information relatable, and can implement it and train staff on its proper use.

2. **Give access points their own security.**

   Network data points and wireless access points fall under the "access points" category. It's critical for these things to have their own security. Consider these questions:

   - Can someone connect a laptop to your network and access data?

   - Is your wireless access protected?

   - Do you allow wireless guest access? If yes, is it segmented from your business network?

   - Is your network infrastructure protected? Are your switches, servers, routers, and other network devices secured behind locked doors, even in locked cabinets?

   - Can USB flash drives work on your network? And should they?

   - What new mobile devices, such as smart watches, pose new threats?

   The questions are endless and most business owners do not know the answers, unless they are an IT firm. We'd like to leave you with this one last question: Doesn't it make sense to save your energy for business development and leave the complicated, ever-changing detailed IT questions for those who specialize in that area? As the saying goes, you don't go to the doctor for car maintenance. Likewise, you don't trust amateurs with IT.

3. Close unprotected operating systems.

One of the most important things that a business that wishes to stay in business will do is ensure that their operating system and software programs are current. This cannot be emphasized enough! Many individuals believe they can do this effectively on their personal technologies, and they likely can. However, a business requires an entirely different approach. You can't put off updates and specific actions that make you less vulnerable.

Does it make more sense to know that these things are being done right when the patch (that's what removes glitches from programs) is released, or when Stan in accounting gets a few extra minutes. . . even if it's a month out? Immediate is always better with updates to operating systems. IT specialists get these patches right away and have the resources to apply them quickly. Things you need to know: Are Java and Flash updated automatically or manually? Can your business really schedule in immediate times for updates with internal staff, as compared to external IT help? And then there is this consideration: Does your organization run programs that will fail if certain updates are applied? Hopefully not!

4. More than just "antivirus" protection must be used to protect a system.

Antivirus programs remain a necessary component to overall security, but there are Spyware and PUPs (Potentially Unwanted Programs) that know how to sidestep nearly every antivirus program out there. This is how cybercriminals stay in business. If you're wondering how you may get a PUP in your system, you may be shocked to find that you likely invited it in, usually by downloading something—maybe a game, some social media app, and even some programs such as free business utilities. When we inform people of this, they usually ask, "What should we do with those programs?" Our answer is always, "Get rid of them!"

One of the best forms of protection from spyware and other Internet threats is OpenDNS. From experience, we believe so strongly in this product that we are glad to mention it by name. Its value has proven itself time and again for our customers.

How does OpenDNS work? DNS (Domain Name Service) converts a common name, such as www.google.com, into the numeric address that runs the Internet. WHAT? It's far easier for users to remember names than numbers. Therefore, when a user requests access to a website, the DNS system converts the common, readable hostname so that the request can be completed. Furthermore, it creates an incredible database of categorized sites. All of this works in your favor to help curtail malware from making its way into your system. Another benefit of OpenDNS is how it can filter content to provide protection against inappropriate data being displayed in a public area, such as a café or waiting room.

5. **Stop illegitimate emails from ever reaching your inbox.**
Cloud-based filtering solutions can keep the junk out and also provide a portal for archived email storage, as well as act as an emergency portal if your mail server is offline for some reason. Both inbound and outbound filtering should be employed. Inbound for reasons that we hope are obvious by now, and outbound filtering to protect your customers and vendors from getting spammed— just in case something does get loose on your network. Don't innocently pass it on! Outbound filtering can also serve as an alert to a possible issue, should items being sent get quarantined.

6. **Business continuity.**
Remember, that IT is ongoing—not once and done. Is your data backed up both locally and offsite? How often are backups tested? Can you restore in quick order? How often do you try it by having a 'dress rehearsal' for the real thing? Your readily accessible and usable backup is essential to your security as it provides business continuity.

MAKE YOUR DECISION BEFORE A CYBERCRIMINAL MAKES IT FOR YOU.

Threats may come in many forms, but so does protection.

Failing to take security seriously is a larger business risk than most businesses would knowingly take. All your hard work and endless hours were not spent so someone could jeopardize your livelihood because they were able to enter your system and find something of value, while leaving you with a failing business. If you don't want to start all over again, you want to do what you can to protect your data. There is no other option. Even if you have insurance to help recover from the loss, it will do little to help restore customer confidence in you or restore your reputation.

If you don't have a relationship with an IT professional right now, take the time today to create one. **Every day you waste is a day that you are vulnerable.** Use this chapter to show your IT professional what you want addressed, ask questions, and find a way to make it work.

About Daniel

Daniel Foote has over 15 years of experience in the IT industry. From being the sole support for a small business network to managing a support desk that delivers services to thousands of customers through all methods of data delivery, Dan has a wealth of experience on many platforms.

Currently, Daniel is the president of **DanTech Services, Inc.**, an IT Managed Service Provider based in Anchorage, Alaska. DanTech Services works with small-to-medium sized businesses to protect their technology infrastructure, data, and users with a layered approach to security. The DanTech Services team provides support from the desktop to the cloud, manages web servers running the latest Content Management Systems, delivers perimeter security, and protects against data loss with intelligent business continuity services.

Dan is a featured presenter for the Anchorage Small Business Development Center (SBDC) and other venues covering topics such as cyber security, business continuity, and cloud computing options. Taking away the business owner/manager's anxiety is a major deliverable that DanTech Services provides its customers.

CHAPTER 25

DEFENDING YOUR BUSINESS AGAINST A DATA BREACH: BE PROACTIVE, NOT REACTIVE

BY RICHARD RAUE

USA Today reported in September of 2014 that **43%** of companies had experienced a data breach within the past year. That is a staggering number, isn't it? From my professional view, this percentage can be looked at in two different ways. If you are a consumer, you likely think, *nearly half of all businesses I visit may experience a data breach—ouch!* And if you're a business owner, you think, *how would I ever recover from a data breach and come out of it with my business intact?* Both are valid concerns and **it is the responsibility of business owners to find solutions that help ensure they stay out of that 43%**. In fact, it's in everybody's interest, aside from the cybercriminals who are looking for unlocked virtual doors that give them an invitation, to access a business' information.

DATA BREACH LAWS

If it's presumed lost, it's presumed breached.

California enacted the first data breach laws back in 2002 to start protecting consumers in an ever-growing online world. The law required:

A state agency, or a person or business that conducts business in California, that owns or licenses computerized data that includes personal information, as defined, to disclose in specified ways, any breach of security of the data, as defined, to any resident of California whose unencrypted personal information, was or is reasonably believed to have been, acquired by an unauthorized person.

As you read, "as defined" is a key component to this type of legislation, but it's not quite as subjective as you may think. *It includes everything that could be used to help a cybercriminal take your information and use it for their own gain, or malicious intent.*

The law that was put into place in California has been the model for most states' data security laws. There are a few states that are stricter, such as Massachusetts, which doesn't even allow businesses to ask for a driver's license number for a check. Why? Because that license can also give someone access to you. If a driver's license number is sent in an email, it's a breach. Very direct and easy to assess. To date, every state has passed their own data breach laws, aside from three, which are: Alabama, New Mexico, and South Dakota. (Note: Alabama and New Mexico have introduced legislation in 2015.) No information can be found to answer why they haven't passed these laws yet; however, consumers who live in these states need to realize that there is no law to state that they must be notified in case of a breach that isn't covered by a federal act or regulation (for example: HIPAA or HITECH Act). For consumers in these states, they could be notified if a breach happens, but it is because of the decision of the business to disclose it.

Many states have data breach laws based off of California's pioneering legislation as a standard. To give you an idea about data breach laws, according to Louisiana Revised Statutes 51:3073:

"Personal information" means an individual's first name or first initial and last name in combination with any one or more of the following data elements, when the name or the data element is not encrypted or redacted: (i) Social security number, (ii) Driver's license number, (iii) Account number, credit or debit card number, in combination with any required security code, access code, or password that would permit access to an individual's financial account.

This type of data breach verbiage applies to everyone, which is important to know. *And the laws are applicable to all types of businesses, including: healthcare, financial, retail, insurance, churches, and schools.* The medical field has some of the most stringent acts out there. HIPAA (Health Insurance Portability and Affordability Act) and

HITECH (Health Information Technology for Economic and Clinical Health) Act are very detailed acts to protect consumers. Businesses that fall under these acts that do not comply will find themselves facing fines, and even possible criminal implications.

Why are medical records so valuable? Medical offices have almost a complete consumer profile, making them appealing and profitable for cybercriminals. You have access to names, birthdates, insurance information, social security numbers, addresses, driver's license numbers, medical history, medications, employer, phone numbers, and everything that it takes for someone else to become you. It could be for financing addictions, or turning a quick profit from purchasing things on your credit card.

This is overwhelming and startling to most. *How do you know when your information is in the wrong hands*? You either get the shock of your life when authorities knock on your door and tell you, or else you get a notification from a business that they've been compromised, which leaves you in an equally compromising situation!

DETERMINING WHEN A DATA BREACH HAS OCCURRED

Only data that is encrypted is considered not breached.

Many people immediately think of credit card information when they hear data breach. It's a significant part of it, but not the largest motivator for cybercriminals. *According to IDTheft.org, in 2014 the top three breaches occurred in the following industries*:

1. The medical/healthcare industry: which experienced a staggering 42.5% of breaches. Medical records are golden on the dark web (the Internet's black market), because they contain so much information.

2. The business sector: which experienced a 33% data breach rate.

3. The government/military sector: which had an 11.7% data breach rate.

All of these numbers are very alarming and they show that despite business' efforts and government acts and laws, we all remain vulnerable. One business failing to protect our data can lead to many agonizing hours of frustration trying to clean up the mess. It's no wonder that approximately 60% of businesses fail after a data breach. It's costly to repair and even if you make amends with those who have been breached, you seldom regain your reputation or trust.

Most data breaches are not immediate attention getters; they are slow infiltrations that take place over time. One of the most infamous stories that inspired stronger laws and acts about data breach happened with ChoicePoint, an actual data aggregation company, who became compromised in 2005. *Just imagine…*

The CEO and COO sold $16.6 million in stock the day before they revealed that 160,000 consumer records were stolen.
Furthermore…there were no charges filed!

Today, individuals and the media must be notified and CEOs and COOs have a responsibility to everyone, not just their pocket books' well being. Through the example of ChoicePoint, we learned a lot about how people "in the know" were able to protect themselves while a great many other peoples' lives were given an instant dose of stress. . . and lingering questions: *Who has my information, and what are they doing with it?*

Thankfully, HIPAA laws require that consumers, the Secretary of Health and Human Services, and other impacted parties be notified immediately upon discovery of a breach. **Every day is a valuable day when you're trying to stop your private data from working against you**. Yes, there are immediate solutions such as credit monitoring offered to help impacted consumers by some businesses, but that business has to be big enough to afford that cost. And remember, small businesses are huge targets for cybercriminals, because they often have the least amount of IT security measures in place.

Consumers are at the mercy of businesses to have the utmost integrity in protecting their confidential data, making it only available for its intended purpose. This means that businesses must embrace *prevention* and keep in mind that the cost of a breach is considerably larger than the cost of having the proper IT in place.

STRATEGIC AND EFFECTIVE SOLUTIONS
TO PREVENT DATA BREACHES

Typically, IT pros are overly confident about their "know how" for preventing data breaches.

When any business lets their guard down and just assumes that everything is in place with their IT that protects the business' interest, disasters are more likely to happen. This is evidenced by the PC World report that 27% of companies do not even have a data breach response plan or team in place. This is down from the previous 39% statistic, but we still have a long way to go. Each and every business should sense the importance of this—from the alarming number of stories on the news if nothing else. **They should also stop believing the myth that their information is not valuable to a hacker, because it is**. Every year cybercriminals are becoming craftier and more determined, working to get at information and doing everything they can to stay one step ahead of the hardware and software glitches that help them gain access to information.

The key things for any business to keep in mind when they are creating and implementing a data breach plan is that:

- The plan must be effective. This seems logical, but it's necessary to relay. As of today, recent employee surveys on the topic of "confidence in their employer's data breach plan," show that a mere 30% of employees interviewed believe that the plan that's in place is truly effective. This number should be 100%. As a business owner, you should want to know this information and see if employees are seeing something you possibly are not, or that their IT person isn't. Data security isn't something that your average IT guy is going to be able to master. It takes more in-depth knowledge than just software updates and data back-up.

- The plan must be evaluated. Quarterly evaluations are necessary, because cybercrime is always progressing and moving forward, finding new ways to keep doing what it does best—wreaking havoc on others. It's estimated that only 3% of businesses do quarterly evaluations of their data breach plan—which is 97% too few. According to Ted Julian, who is the Chief Marketing Officer with Co3 Systems, which handles cyber-incident response management, says, "Most organizations, and I'm only talking the sophisticated ones, have done a little, but it's not enough."

- <u>Only data that is encrypted is considered not breached</u>. There are no exceptions to this, which means that encryption is something key in a business' efforts to significantly reduce the risk of a breach occurring on its system, or a hack not being called a breach.

- <u>Have scheduled and regular risk assessments</u>. Not only is this a requirement of the HITECH Act, but it is a sound strategy to make sure that a business positions itself more favorably in a world where breaches will continue to be attempted on all businesses.

- <u>Maintain all systems</u>. Ensure that patches, updates, and antivirus software are all up to date all of the time. There cannot be exceptions to this, which is why the most protected businesses use professional IT services to handle these tasks. They want to reduce their vulnerability as much as possible.

- <u>Do continual audits to firewall logs</u>. This is the most effective way to determine if there is any "unusual activity" taking place, or something that doesn't really align with the standard activity in a business. This is a great way to catch an early sign of a possible breach, or at least alert a business that someone may be trying to infiltrate their system. Remember, malicious-minded people do not quit on their goals to get information—it's their livelihood, but could be the end of yours if you are a business owner who doesn't take it seriously.

Businesses who embrace the "it'll never happen to me" mentality with their data security—regardless of size—are putting a message out there in the virtual world that says, "Come on in, we left the door open." *The right IT personnel ensuring a business has its best chance against cybercriminals sends a very different message, "I'm right here and I'm watching. Move along."*

IS YOUR DEFENSE SYSTEM IN PLACE?

*If a business attempts to hide a data breach, they will
face increased penalties and jeopardize their business being
able to rebound from the disaster.*

Author David S. Wall penned, "In a nutshell, we are shocked by cybercrime, but also expect to be shocked by it because we expect it to be there, but—confusingly—we appear to be shocked if we are not shocked (if we don't find it)!" Confusing? Yes! However, it really sums it up perfectly for many business owners approach to their data and how they perceive its value. **We know cybercrime exists and we don't want to be impacted by it, but we don't invest as much effort into protecting our livelihoods and our customers from it as we should**.

Knowing that anyone who is connected to the Internet is at risk of being hacked, *it's important to acknowledge that data is valuable*. Hackers will go after it no matter how secure a business network is. The difference for who gets "hit" and who doesn't comes down to one thing—**defense**. A good defense against crime must be set up in a business' virtual world so they can always know what is happening and respond more quickly if a criminal finds a way in. *No matter how many systems are in place, you need to pay very human attention to the data that's being offered to you.*

Time and data demonstrates that regardless of acts such as HIPAA and HITECH, we still are in an uphill battle against data breaches. When an act mentions a business tending to what is "addressable," addressable is not an option or a suggestion. It is an expectation, one that could cost a business dearly if they don't comply with it. The penalties are expensive and the cost of encryption will always be less than the loss a business faces from a data breach—it's a loss of consumer trust, money, and possibly the end of that business. The odds aren't in the favor of any business surviving a breach.

The stories and compelling cases of why a business is taking its greatest risk if it doesn't proactively address data breach prevention are endless. You can find them everywhere, and hear them on the news far too often. As a business owner, you've invested a lot into creating a business to be proud of, as well as one that is profitable. *Why risk that to hope that you don't fall into that 43% of businesses who experience a data breach?* We are living in an age where consumers are aware of what is happening with their data and just because a business may want to ignore the harsh reality of cybercrime, consumers become more savvy about it—especially if they've been a victim of it. *They will gravitate toward the prepared businesses, the ones that value their customer's data through their actions and strategies, not just through their words.*

Be the business that they want to confide in!

About Richard

Richard Raue, CHSP, CHSA is the founder and principal owner of **Hi-Tech Computers of Ruston, Inc.** He started Hi-Tech Computers in 1997 while finishing a Computer Science degree from Louisiana Tech University in Ruston, Louisiana.

Richard's initial focus was helping small businesses and residential clients manage their technology. Hi-Tech steadily evolved into a healthcare-focused MSP practice primarily working with medical clinics and rural hospitals. Recognizing the need for ever tighter security and seeing clients struggle with the complexity of Healthcare IT regulations, Richard and his team have invested heavily into certifications and training for every single staff member. Hi-Tech proudly claims to have a 100% HIPAA trained workforce. Richard holds certifications as Certified HIT Security Professional (CHSP) and Certified HIT Security Administrator (CHSA) along with many industry-specific certifications covering various hardware and software solutions.

When not at Hi-Tech, Richard is heavily involved in the community. As a member of Kiwanis International since 1999 and having held positions as Director, Club President and Lt. Governor for Louisiana-Mississippi-West Tennessee, Richard has participated in many service projects that directly or indirectly affected hundreds of local children. Richard also served as a director on the board of the Ruston-Lincoln Chamber of Commerce for five years and is an Advisory Council Member at NewTech @ Ruston High School.

Having five boys of his own, Richard is an Assistant Scout Master for Boy Scout Troop 59 and teaches Fishing, Wildlife Management and Communication merit badges.

CHAPTER 26

SECURITY CONSIDERATIONS FOR FINANCIAL PROFESSIONALS

BY ROBERT BOLES

THE FBI COMES CALLING

It was a Monday morning, and John L. arrived at work to an urgent message from the FBI regarding some suspicious activity which required John L.'s cooperation. There had been some irregular trading activity of the company's stock, and the FBI was investigating. The agent had insisted that the matter be confidential, and that John L. provide any insight possible via the existing logging and security measures in place. Unfortunately, John L.'s company, like many others, had been lax in maintaining sufficient security and now it was an issue.

"Let me get to the point, Mr. L.," said the agent. "We have observed some irregularities potentially coming from your network or one of your users, and because you are regulated by the government, we need to investigate. We have identified anomalies in recent trading activity and we suspect one of your executive's laptops may have been compromised, and potentially exposed your company's confidential data."

The agent continued describing the penalties involved due to a potential security breach. The implications hit John L. like a ton of bricks. The company's reputation and trust could be severely damaged. Additionally, the loss of key intellectual property and client information could be devastating. John L. thought to himself, "How could I have missed something like this going on in my office on our server?"

"We need to locate the data leak and remediate this situation. We are asking for your full cooperation. We need to access your logs and reports," added the agent.

"Can you tell me what we are looking for or who you suspect?" asked John L.

"At this point in the investigation, we cannot disclose that information," the agent responded.

THE BIGGER ISSUE

John L. was concerned about this potentially devastating news for the company. A potential security breach had occurred and he had no ability or insight into if, how, or even when this might have happened. Even worse, was it still being committed? And how widespread could it be? What liability did he and the company have?

The FBI left without much more information to share other than there was activity under suspicion. John L. needed to know what was really happening on his network. He called our company in to assist.

"Listen Rob, I need some help fast. The FBI is breathing down my neck, and they are trying to determine if there is foul play around one of our executive's activity. I don't have the resources to track this and discover what is happening on our servers. Can you help me out? I need you to be discreet, as I could be in trouble if I step on the FBI's toes. They want me to provide insight into where this could be coming from. In addition, we want you to help us identify how to better protect our systems so something like this can never happen again."

TRACKING AND LOCKING IT DOWN

We showed up at John L's company after hours and put our tools to work on his network. Within 90 minutes, we could quietly observe in the background without tipping anyone off. We had full visibility of all traffic and began identifying anything suspicious. For the next couple weeks, we worked cooperatively – identifying activity on his company's network. Fortunately for John L., the issues we identified were not issues the FBI investigation was looking for. However, we were able to identify a company executive's Cloud-hosted email account that had been compromised by someone outside the company. John L. breathed a sigh of relief that the breach was not on his internal network. This remained serious news, but the confidential client data was intact and a more secure means of authentication would be implemented.

LESSON LEARNED

While this could have been a catastrophic event for the company we assessed, the underlying task was to prevent it from happening in the future. The first thing we discovered was that John L.'s company had a general security and employee use policy on computer and Internet use, but implementation was not aligned with that policy. The company assumed everyone was doing what they were supposed to be doing.

"Its good that you have a policy John L., though if the network security is not aligned with the companies documented policy, or if it is not being enforced, its not much different than having no policy," I explained. We went to work.

COMPLIANCE IN THE FINANCIAL SPHERE

Daily we hear about another big company being hacked. In the past five years, literally millions of personal records, credit cards, health records and bank accounts information have been compromised. The list includes Adobe, Anthem, Target, JPMorgan Chase, Citigroup, TD Ameritrade, and this information then becomes available for purchase on the Darknet from "hacker shopping malls" where all sorts of data, hacking tools and even hacker "helpdesk and support" are available 24 hours per day, 7 days per week to those who seek it.

Furthermore, it's not just big business which is being compromised. Small and medium-sized business are easy targets too. Businesses of all sizes benefit from a strong strategic IT plan. Insurance companies, stock brokerages, investment companies and money manager firms also have to be concerned with compliance because of governmental regulation. Compliance regulations like Sarbanes–Oxley, Gramm–Leach–Bliley, and PCI-DSS are not difficult to meet as long as there are a set of best practice models and processes in place.

THE WRITTEN POLICIES

Like we do for many other clients, we assisted in drafting policies that worked for John L.'s company. Upon completion, we encourage our clients to have an attorney review and approve their new policies.

Once the security policy is complete, we translate it into rules and access privileges on the network which align with the company's written policy. This becomes the standard for computer and technology use throughout the entire organization.

SECURITY IN PRACTICE

Security is not a "set-it-and-forget-it" aspect of business. A strong IT strategy with security as core to the overall process frequently follows a process of Assess, Plan, Integrate, Evaluate, Maintain and repeat. This is a necessity as IT is constantly changing and evolving. Clients needs are constantly changing as well. Security cannot be an afterthought or "bolt on" component, nor is security simply a box to check off in an audit.

A resilient IT Security strategy combines hardware, software and processes. An element of process is identifying security components that communicate and work together to protect your users and business assets. Time after time we have completed assessments of a client's environment and have identified various security platforms and tools. Each of these provides a specific security function, but because they do not communicate and operate as a cohesive defense system, they are not completely effective and leave gaps. Security in practice requires a complex skillset applied by highly-trained expert personnel. Often human error and misconfiguration are the root cause of a host of security issues.

MANAGED SECURITY SERVICE PROVIDERS

Many companies are resistant to hiring and creating a whole department that maintains and oversees the security of their network. IT Security talent is in demand and the tools and methodologies are evolving rapidly. Keeping up with the security needs of business while maintaining expertise and infrastructure to support and run the business is expensive.

The good news is that all your security monitoring can be accomplished remotely. Companies like ours specialize in this type of cyber-security. We serve many clients and are on the front lines battling the latest threats. We have specialized expertise, tools and processes to protect you and your business at cost effective rates. The technology available for defense in the security space is accelerating ridiculously fast – in both capabilities and the completeness of defense that we are able to provide. This is combined with teams which work 24x7x365 as extensions of client IT departments and it is a constructive partnership.

The security solutions we put in place identify all traffic on our client networks by application and provide full visibility to all data traversing the network. This is all made available to clients via a graphical summary report of the applications, users, websites accessed, threats and content on the network. These solutions can be integrated with Active Directory, which is the authentication mechanism controlling how users access the network and what resources users are able to access. By leveraging a security solution that provides visibility to all data traversing the network, including endpoints and mobility/BYOD, we are able to assemble a more complete picture. With visibility to all the activity by application and type, we are able to identify what users do on a daily basis, what websites they are trying to go to, how long they were there, what resources they are using internally, and if they are trying to access things they are not supposed to. The security solution is able to enforce the written security policy on a very granular scale. Furthermore, if a user is not recognized, they are not permitted on the network.

Now let's talk about Zero Day Threats. Traditional Anti-Virus tends to be two months behind in blocking threats. The most complete and sophisticated security solutions now block Zero-Day and unknown threats. With the proliferation of ransomware and threats such as Cryptolocker (which encrypts all your data and holds it until you pay the ransom), being able to recognize potentially malicious content is more critical than ever. The solution includes dynamically analyzing suspicious content in a cloud-based environment where it can be executed in a safe place and its behavior observed and a determination can be made if the threat is malicious or harmless. If malicious, the malware is blocked at the firewall and all activity is documented in a report and made for reference.

With these tools, financial companies have the ally they need to comply with oversight by entities such as the Security and Exchange Commission and Financial Industry Regulatory Authority.

MERGING COMPLIANCE WITH BEST PRACTICE

Governmental oversight requires a certain level of security. It is up to each company to know what these regulations include for meeting specified requirements as part of yearly audits. If you apply industry expertise and best practice, including written policies, compliance issues are much simpler to mitigate.

In John L.'s company, they did have a partial policy in place. Though the vulnerability was ultimately "outside" the CORE network, the incident showed they were not in total compliance. Fortunately, no user data, client data or intellectual property was compromised, though this incident became a very expensive lesson for John L.'s company. There were substantial fines for being out of compliance.

TOP TWELVE ACTIONS YOU CAN DO TODAY

1. *Have an attorney-approved written security and employee use policy.*
   This policy will provide the roadmap your business Information Technology team will use to adopt and integrate security solutions and actions into day-to-day operations.

2. *Follow the policies you have in place.*
   Having a policy and following it only nine times out of ten is not compliant. When regulators are auditing, the one time you are not in compliance is what will be recognized. Maintain consistent documentation to validate and support these actions.

3. *Schedule regular DR testing.*
   The time to find the holes in the process is not during an actual event. Testing your DR process helps identify gaps and solves them, creating confidence in and validating the plan. It is an opportunity to demonstrate operational readiness in the unfortunate event it is required.

4. *Schedule regular Security Audits.*
   As part of the security audit process, validate configuration of systems, including active accounts, usernames, passwords, user groups and access, Network and Security Assets.

5. *Limit User Privileges.*
Establish who has access to available data and the ability to install applications on computers.

6. *Filter Spam/Malicious emails in the Cloud.*
Do not allow SPAM and Malicious content on your network. There are many mature and robust cloud-based filtering options which keep SPAM and Malware out of your email and integrate with URL click protection to keep users off malicious websites and free up your bandwidth. This is a 'no brainer.'

7. *Maintain best practice update cycles.*
Unpatched workstations are gateways for the bad guys to access your network.

8. *Have an incident response plan.*
In the event something occurs, and it will, have a plan. Many companies are unable to react to events due to poor incident response plans. Often companies do not even have an inventory of their data.

9. *Ensure that your archiving systems are functioning properly.*
Run searches at least once a quarter to ensure integrity of your data.

10. *Test your backup systems monthly for reliability and data integrity.*

11. *Take inventory of your existing Security Solution and the completeness of vision and protection it provides, and how its individual components communicate and work together cooperatively.*

12. *Stop believing it won't happen to you.*

About Robert

Robert Boles is the Co-Founder and President of **BLOKWORX**. Robert brings more than fifteen years of solving real business challenges for clients, ranging from individual and family-owned to Fortune 500 corporations. Robert's core belief in Security, Reliability, and *delivering a positive user experience* drives BLOKWORX's fundamental practice of building secure and scalable IT Solutions that users enjoy working with.

Prior to founding BLOKWORX, Robert was an Enterprise Systems Engineer for an international communications company providing Advanced IP services based on Cisco, Juniper, HP, and Checkpoint, among others. Robert was a highly-specialized engineer, one of only four in a company with over 200,000 employees. Robert worked cooperatively with Product Marketing, Sales Organizations, and Operations to create a suite of managed service offerings, and defined many of the processes and technology selected within the projects.

As a Systems Engineer directly supporting the sales organizations, Robert designed, successfully implemented, and seamlessly transitioned to the support team over 100 managed network solutions for clients among over 1,400 global locations. As a subject matter expert, Robert educated clients, engineering and sales teams on the Advanced IP product portfolio, which consisted of Wide Area Network Services, Network Security Services, Colocation and Advanced Hosting.

After seven years with the company, Robert made the decision to go back to his roots, and create an IT service provider which met his vision. BLOKWORX was driven from the start to solve real business challenges with technology – by delivering secure, scalable IT Solutions for its clients. Given BLOKWORX's exceptional client retention and long-term relationships, the formula is working.

Robert was raised in Arbuckle, CA, a rural, agricultural town outside Sacramento. His father was a local businessman, and his mother managed the local bank. Following high school, Robert joined the United States Marine Corps and served in Operation Desert Shield/Desert Storm. This early-life experience impressed upon Robert the value of relationships, doing right by people, and teamwork. These were lessons which have fueled the BLOKWORX philosophy of: *If we always take care of our clients by doing right by them in the most honest way, and have fun in the process, then everything else will take care of itself.*

BLOKWORX is a member of the National Veteran Owned Business Association (NaVOBA), and proudly supports CompTia Troops to Tech.

A featured speaker for both internal and client presentations and events, Robert is widely valued for his integrity, passion for technology, and ability to solve business challenges.

Robert lives in San Francisco, CA with his wife Sarah, son Jack, and dog Chewie.

More information available at:
www.blokworx.com
http://www.troopstotechcareers.org
Email: robert@blokworx.com
Office: 415-571-4300

CHAPTER 27

THE RISE OF MSSPS – MANAGED SECURITY SERVICE PROVIDERS

BY STUART SANDERS

INTRODUCTION

Many industries go through phases of growth including early tinkerers, commercialization, industrialization and then "utility-isation". Take electricity – once upon a time, any business that wanted electricity would build their own power generation plant, but economies of scale led to electricity as a utility. It's the same with transportation networks and telecommunications and many others. Even a car manufacturer does not make every component in a car.

Cyber security has gone through similar phases. First, it was the tinkerers that played with viruses and other threats. Many of these became commercial interests, including McAfee and Kaspersky. This chapter will discuss the history of cyber security threats, the rise of specialist, cyber security providers, and the benefits of using a cyber security provider.

EARLY VIRUSES AND MALWARE

Computer viruses have existed much longer than most people realize. The concept was first discussed in 1949 by John von Neumann at a University of Illinois lecture on self-replicating computer programs and later published in a 1966 essay titled "Theory of self-reproducing automata". In the 1970s, a self-replicating virus called *Creeper* was released on Arpanet (the predecessor to the Internet). The *Creeper* was an experimental piece of software.

The first "in the wild" virus to infect personal computers was a piece of software called the *Elk Cloner* in 1982, which infected Apple computers using their DOS boot disk. Thus began the ever-increasing complexity of viruses and varied targets. *Brain* infected MS Dos computers in 1986. *Winvir* infected windows 3.0 in 1992. *Vlad* infected win95 in 1996. Other operating systems were even targeted, such as the Amiga home-gaming system which was infected with the *SCA* virus in 1987.

These early activities kick-started the antivirus industry and lead to the creation of companies such as McAfee, which used a signature-based system to detect and remove known viruses. From there, an arms race began. Malware grew increasingly more sophisticated, from encrypted or packed files that make signatures hard to create, to polymorphic viruses where code is actually rearranged, to viruses that attack known antivirus applications and delete or damage them, and to stealth viruses that dig deep tendrils in the operating system to intercept attempts to find them.

With each advance, antivirus companies had to respond and develop countermeasures. In the early days there was no Internet. Viruses infected through floppy disks, bulletin board systems (like AOL), and even legitimate software disks, because the developer had an unknown infection.

Then the Internet came along. Viruses began to expand much faster. New types were created, such as macro-viruses that sent emails with infected documents across multiple operating systems and the infamous *I Love You* virus in May 2000.

Viruses also targeted specific software that could be exposed on the Internet, such as the 2003 worm *SQL Slammer*. This malware remains the fastest spreading virus/worm of all time. It infected all exposed systems on the Internet within fifteen minutes and crashed the Internet due to the overwhelming traffic it generated.

In the early 2000's, most people still did not use firewalls on their personal computers, and even most small companies connected computers or servers directly to the Internet. Early personal firewalls were complex and required knowledge most people lacked. Only companies with corporate IT staff or external consultants installed and used firewalls. Even if a firewall came with a computer it was not usually turned on because it would block Internet traffic.

This changed in August 2004 when Microsoft released Windows XP service Pack 2 in reaction to industry pressure to address the issue of malware and software bugs. This major update included a firewall with Windows and *turned it on by default*. And so the march continued. Malware continued to get more sophisticated and the security industry continued to react. And then something changed...

ADVANCING MALWARE AND VIRUS THREATS

Up until this point malware continued to be developed mostly by hackers who wrote viruses for kicks. Then money became involved. The initial dot-com era had passed, and hundreds of millions of people were now online – surfing the Internet, exchanging emails, and spending money on Amazon and other popular sites. The bad guys now had the means to make money. Credit card information could be stolen and used to make online purchases. Banks introduced online banking and accounts were now "available" to be hacked and money transferred out.

Perhaps the biggest threat to emerge is the rise of the **botnet**. An infected computer can become part of an army of slaved computers controlled to send spam or viruses, attack websites, blackmail ecommerce systems, serve porn or pirated and other undesirable content, and steal data and identity information from the infected machine, including banking information.

Recently an underground industry has emerged to provide hacking-for-hire, stolen data with paid zero-day exploits[1], botnets, denial of service attacks for hire, and much more. This underground industry is known to provide customer service levels you cannot get from most legitimate software vendors, including 24/7 support hotlines and guarantees on profitability.

---

1. A zero-day (also known as zero-hour or 0-day) vulnerability is an undisclosed and uncorrected computer application vulnerability that could be exploited to adversely affect computer programs, data, additional computers or a network. It is known as a "zero-day" because once a flaw becomes known, the programmer or developer has zero days to fix it. (Wikipedia)

VIRUS AND MALWARE ACTIVITY IN RECENT YEARS

In today's world, Cyber Security issues are often a feature on major news. There have been a rash of major security breaches since the end of 2013. Many major companies have been publicly hacked or experienced data breaches of magnitudes never before seen.

In November and December of 2013, hackers were able to acquire 40 million credit and debit card numbers along with 70 million records of Target customers. (Perez, 2014) The incident stayed in the news for several weeks, as investigators discovered the breach encompassed a much bigger data set than originally suspected. JP Morgan was breached in the summer of 2014, and data affecting 76 million Americans and seven million small businesses was stolen. (Jessica Silver-Greenberg, 2014) September of 2014 brought us news of the Home Depot breach, where details for 56 million credit cards were stolen. In October of 2014, hackers infiltrated Sony Pictures' network and servers, stealing and publishing terabytes of confidential documents including unreleased movies, details on film productions and employee data (Peterson, 2014).

Then the Anthem breach occurred in the beginning of 2015. 79 million insured customer details were stolen from Anthem. In this breach, criminals got names, birthdays, email addresses, Social Security numbers, medical identification numbers, addresses, and employment data, including income. (Weise, 2015) Together, this data provided all of the information necessary to impersonate a victim, setup bank accounts, apply for credit cards, and more.

The financial cost of these breaches is staggering. The Ponemon Institute estimated "the cost of data breaches due to malicious or criminal attacks has increased from an average of $159 to $174 per record." (Ponemon Institute, 2015)

CYBER SECURITY RESPONSE

Many small companies, and some not-so-small ones, still believe cyber security simply consists of installing some antivirus software and a firewall. Once upon a time, this was not only sufficient but put you at the forefront of protection. However, the security landscape has changed quickly in recent years. It is said that today cyber security defenders need to get the job right 100% of the time, because bad guys only need to be good *once* to get in and own your network.

Many in the cyber security industry are moving away from the belief that the edge of the corporate network is like a wall and now assume something will get through. Detection and remediation is now key. It is not to say that the edge of the network should no longer remain a bastion of firewalls and other security services, but it is widely becoming recognized as simply not enough.

The Gartner report "Five Styles of Advanced Threat Defense", (D'Hoinne, 2013) listed the following five styles of advanced threat defense:

1. *Network traffic analysis*: monitoring network traffic, establishing a base line, and looking for anomalous patterns.

2. *Payload analysis*: use of a sandbox, either locally or remotely (cloud), to observe an open payload (script, app, etc.) and detect wrongdoing.

3. *End point behavior analysis*: a range of methods including isolating applications in virtual containers and blocking attacks by monitoring system configuration, memory, and running processes.

These three styles are real time (or very near) protection styles requiring split-second decisions be made on whether to pass or block traffic, payloads, or specific behaviors. Antivirus and firewalls form one small part of these types of threat defense.

A few post-compromise styles also exist:

4. *Network forensics*: capturing and keeping all network traffic for later analysis, requiring analysis and replay of incidents to allow an organization to rapidly respond to an incident, determine what occurred and how, close any potential holes, and follow-up with clients, regulators, and partners.

5. *Endpoint forensics*: collecting data from monitored endpoints to assist with incident analysis and response.

Two of the five styles are based around the assumption that a breach will occur. These styles of defense are wide and varied in discipline and involvement. For an organisation utilizing databases, web tools, mobile applications, and multiple servers and endpoint operating systems, you need a skill-set not easily found in a single person. Depending on the size and scope of the organization, a small IT team may not be enough to provide active defense followed by detection and analysis of anomalous or unusual behavior.

The skills required for cyber-defense encompass networks, applications, virtualization, operating systems, data analysis, forensics and the ability to think like an attacker. This is not a simple list. An average IT person can probably adequately configure a firewall and install endpoint security software, but do they understand buffer overruns and process injection? Can they do deep traffic pattern analysis to pull the anomalous traffic aside for further review? Even within the cyber security industry there are few people who are experts in all domains. So how is a company to make sure that they are fully protected?

MANAGED SECURITY SERVICE PROVIDER (MSSP)

Enlisting the services of a Managed Security Service Provider (MSSP) is the best solution to adequately protect your business from the gamut of ever-evolving security threats. MSSPs are companies that specialize in one domain - managing and providing security to their customers. MSSPs provide differing levels of security, from simply managing a firewall and using ITIL[2] or some form of change management standards to providing a complete security solution including all hardware and systems.

A MSSP can attract and retain talented employees, who are becoming increasingly hard to find and afford, as they have access to technologies and clients who are prepared to pay for a specialist. The training of cyber defenders has failed to keep up with supply, creating a shortage of talent which has driven the price too high for most companies to pay for this level of expertise in-house.

COSTS OF NOT USING A MSSP

Recent advances in malware hold corporate and private data to ransom. Today's cyber threats are significantly more sophisticated, using advanced encryption systems that remain unbreakable, command and control systems hidden behind TOR[3] or on a complex network of

---

2. ITIL - Information Technology Infrastructure Library, is a set of practices for IT Service Management (ITSM) that focuses on aligning IT services with the needs of business. (Wikipedia)

3. Tor is free software for enabling anonymous communication. The name is an acronym derived from the original software project name The Onion Router. (Wikipedia)

compromised computers, and even global networks of criminals and mules in order to evade capture.

Most businesses are doing the minimum to address security. Executives and business owners need to dedicate more resources to cyber security. The risks of not taking of not doing so are numerous. The business costs to small or medium-sized businesses could result in total loss of business. For example, if a local grocery chain or doctor's office was publically known to lose confidential data of community members and that loss of data was due to negligence, or if it was mishandled, customers would look elsewhere for a service provider. The loss of even 20 to 30 percent of customers could send a small-to-medium business into bankruptcy.

Becoming victim to any of today's cyber threats poses a risk beyond just the loss of data, the loss of client confidence is often an even greater problem. The loss to a medium to large businesses can be financially significant but can also cost people their careers. If you have at least done everything possible, such as enlisting an MSSP, then customer confidence will not be as greatly affected. A recent case was when LastPass[4] had a security breach in which one database was possibly exfiltrated. Even though someone probably has a database containing information, because it was properly encrypted using current best practices, the bad actors are unlikely to be able to gain usable information. Properly designed and implemented security systems means that even when a breach takes place the effect is minimal. Customers appreciate the effort, and in the case of LastPass this did not result in a large number of clients leaving.

BENEFITS OF MSSPS

There are nine very tangible benefits to utilizing a MSSP: cost, staffing, skills, facilities, objectivity and independence, security awareness, law enforcement, service results, and technology.

1. **Cost:**
   The cost of an MSSP is less than hiring an in-house team of fulltime cyber security experts. An MSSP spreads the costs out over multiple clients. They benefit from economies of scale because the configuration processes, and advances, can be replicated across the entire client base. Discovery of an issue in one client can quickly be pushed across the client base.

---

4. LastPass is a popular online password management system (www.lastpass.com).

2. Staffing:

The shortage of professionals has put upward pressure on salaries for cyber security professionals, making hiring qualified in-house staff unaffordable. An MSSP can provide career advancement that a small to medium-sized business cannot.

3. Skills:

In small to medium-sized businesses, in-house staff members will only see an occasional incident and often have access to a limited scope of technologies and tools. MSSP staff see a much greater variety of incidents, have access to industry standard tools and leading-edge systems and will possess more relevant experience overall.

4. Facilities:

MSSPs live and breathe security and operate from very secure facilities, passing on the benefits to customers. Access to systems will be engineered to be very difficult and the MSSP may even operate out of hardened Security Operations Centers.

5. Objectivity and Independence:

Many large organizations will have a mix of *adhoc* and non-standard approaches to security across the organization. There may be duplicated or redundant systems or vested interests in some or all of these systems. A MSSP provides an independent and objective approach to standardizing security systems and policies across an organization, removing redundancies and increasing security company-wide.

6. Security Awareness:

MSSPs follow trends, news, and advances in potential threats and vulnerabilities in the security sphere. They have access to advanced warnings from industry peers on new threats and may have access to early release signatures and detection capabilities from vendors. It is hard, if not impossible, for an IT department or a small non-security specialist MSP to do the same.

7. Law enforcement:

If a security breach goes to prosecution, an MSSP likely has experience on the procedures and evidential requirements for successful prosecution. They may also have pre-existing relationships with local, regional, or even global law enforcement agencies.

8. Service Results:

MSSPs typically provide guarantees and service level agreements on service standards, availability, reporting, and even hardware replacement or software uptime. This may include 24/7 availability, a stark contrast to in-house services which only operate during business hours. Attackers are aware of the limitations of in-house security and will often attack on a Friday night with the expectation that a response will not be forthcoming over the weekend.

9. Technology:

The various technology tools used to secure an organization will be more effective in the hands of an MSSP because they are configured and managed by seasoned professionals. A MSSP will be able to respond to an alarm and quickly determine whether it is legitimate threat and needs action taken to block further intrusions.

MSSPS ARE FOR ALL TYPES OF ORGANIZATIONS - BIG AND SMALL

As a small but successful outsourced IT Managed Service Provider we have promoted the use of MSSPs for over 10 years - from the early days of a fledgling industry. We practice what we preach. While I have a deep understanding of cyber security and the risks, we outsource our primary cyber defense to a team of specialists, both internally and with our clients. Do we get involved in the direction and decisions? Definitely! But, the day-to-day monitoring and management of our cyber security, and that of our clients is in the hands of a group of people that do nothing else, MSSP's.

Works Cited

Jessica Silver-Greenberg, M. G. (2014, October 2). *JPMorgan Chase Hacking Affects 76 Million Households*. Retrieved June 4, 2015, from New York Times: http://dealbook.nytimes.com/2014/10/02/jpmorgan-discovers-further-cyber-security-issues/?_php=true&_type=blogs&_r=0

Perez, S. (2014, January 2014). *Target's Data Breach Gets Worse:70 Million Customers had Info Stolen, Including Names, Emails and Phones*. Retrieved June 4, 2015, from Tech Crunch: http://techcrunch.com/2014/01/10/targets-data-breach-gets-worse-70-million-customers-had-info-stolen-including-names-emails-and-phones/

Peterson, A. (2014, December 19). *The Sony Pictures Hack, Explained* . Retrieved June 4, 2015, from The Washington Post: http://www.washingtonpost.com/blogs/the-switch/wp/2014/12/18/the-sony-pictures-hack-explained/

Ponemon Institute. (2015, May 27). *Cost of Data Breach Grows as does Frequency of Attacks*. Retrieved June 4, 2015, from Ponemon Institute: http://www.ponemon.org/blog/cost-of-data-breach-grows-as-does-frequency-of-attacks

Weise, E. (2015, February 4). *Massive Breach at Health Care Company Anthem Inc*. Retrieved June 4, 2015, from USA Today: http://www.usatoday.com/story/tech/2015/07/02/theranos-finger-stick-blood-testing-gets-fda-approval/29625005/

About Stuart

Stuart Sanders is a long time IT aficionado, having first used modern-style personal computers in the early eighties and publishing software through magazines in his teen years. He went on to pursue a degree in Chemical Engineering and an early career in the petrochemical industry. Even through that time, the focus was on process control and computer systems/modelling as this was a natural skill that co-workers, clients and the employers leveraged on.

In 1998, Stuart shifted completely to the IT industry and ran two consulting businesses before joining forces with Velocity in 2011. Velocity is another long-time IT consulting Group in Hong Kong, where Stuart remains to this day as Managing Director of Velocity Technology. Velocity Technology is the IT infrastructure and support business within the Velocity Group of Companies, and has a team in Hong Kong and Manila that provides support to clients across Hong Kong, China, Philippines, Singapore, Thailand, USA and Ireland. Additional support is provided in limited cases to clients with facilities in Vietnam, UK, Australia, New Zealand and Indonesia.

Stuart's experience includes providing IT consulting and strategy for a range of businesses and industries including international groups, but with a heavy focus on the Finance and Financial Services Industries. He has been the primary IT strategist and consultant for a number of multi-billion USD funds and hedge funds and also a listed company on the Hong Kong Stock Exchange main board.